SpringerBriefs in Computer Science

For further volumes:
http://www.springer.com/series/10028

Duy H.N. Nguyen • Tho Le-Ngoc

Wireless Coordinated Multicell Systems

Architectures and Precoding Designs

 Springer

Duy H.N. Nguyen
Department of Electrical
 and Computer Engineering
McGill University
Montréal, QC, Canada

Tho Le-Ngoc
Department of Electrical
 and Computer Engineering
McGill University
Montréal, QC, Canada

ISSN 2191-5768   ISSN 2191-5776 (electronic)
ISBN 978-3-319-06336-2   ISBN 978-3-319-06337-9 (eBook)
DOI 10.1007/978-3-319-06337-9
Springer Cham Heidelberg New York Dordrecht London

Library of Congress Control Number: 2104938018

Printed on acid-free paper

Springer is part of Springer Science+Business Media (www.springer.com)

Preface

In a multicell system, universal frequency reuse can be employed in a network of neighboring cells for higher spectral efficiency. However, universal frequency reuse comes at the price of severe inter-cell interference (ICI), especially at cell-edge mobile stations, which may effectively impair the overall system performance. To actively deal with the ICI, the emerging wireless communication standard advocates the concept of interference-aware multicell coordination. Known as coordinated multipoint transmission/reception (CoMP), the new paradigm allows the multicell system to actively control and even take advantage of the ICI. The objective of this SpringerBrief is to present the architectures of a CoMP system and examine the recent advances in precoding designs for such a system. The motivations and concepts of CoMP are first explored, followed by the review of various CoMP architectures and deploying scenarios in the LTE-Advanced standard. In addition, practical implementation and operational challenges of CoMP are then discussed in detail. Finally, the readers are exposed to latest multiuser precoding designs for the CoMP system under two operating modes: interference aware and interference coordination. Based on optimization theory and game theory, the structures of the multiuser precoders are analytically devised. In addition, the message signaling mechanisms in the CoMP system are systematically developed to facilitate the distributed implementation of the related precoding designs.

The target audience of this informative and practical SpringerBrief is researchers and professionals working in current- and next-generation wireless cellular networks. The content is also valuable for advanced-level students interested in wireless communications and signal processing for communications.

We would like to acknowledge the financial supports from the Natural Sciences Engineering Research Council of Canada and the Fonds de recherche du Québec—Nature et technologies Postdoctoral Fellowship.

Montréal, Canada   Duy H.N. Nguyen
Montréal, Canada   Tho Le-Ngoc
February 2014

Contents

Acronyms

3GPP	Third Generation Partnership Project
4G	Fourth Generation
AWGN	Additive White Gaussian Noise
BC	Broadcast Channel
BD	Block-Diagonalization
bps	Bit per second
BR	Best Response
BS	Base-Station
CQI	Channel Quality Indicator
CDMA	Code-Division Multiple-Access
CoMP	Coordinated Multipoint Transmission/Reception
CSI	Channel State Information
CSI-IM	Channel-State Information-Interference Measurement
CSI-RS	Channel-State Information-Reference Symbol
CSIT	Channel State Information at Transmitter
D.C.	Difference-of-two-convex-functions
dB	Decibel
DPC	Dirty-Paper Coding
DPS	Dynamic Point Selection
FDD	Frequency Division Multiplexing
IA	Interference Aware
IC	Interference Coordination
ICI	Inter-cell Interference
ICIC	Inter-cell Interference Coordination
IEEE	Institute of Electrical and Electronic Engineers
ILA	Iterative Linear Approximation
IP	Internet Protocol
IPN	Inter-cell Interference Plus Noise
IWF	Iterative Water-Filling
JP	Joint Signal Processing
JT	Joint Transmission

KKT	Karush–Kuhn–Tucker
LTE	Long-Term Evolution
LTE-A	Long-Term Evolution-Advanced
MAC	Multiple-Access Channel
MIMO	Multiple-Input Multiple-Output
MISO	Multiple-Input Single-Output
MMSE	Minimum Mean Squared Error
MS	Mobile-Station
MSE	Mean Squared Error
MU	Multiuser
NE	Nash Equilibrium
OFDM	Orthogonal Frequency Division Multiplexing
OFDMA	Orthogonal Frequency Division Multiple Access
PDSCH	Physical Downlink Shared Channel
PMI	Precoding Matrix Index
PSD	Power Spectral Density
QoS	Quality of Service
RI	Rank Indicator
RRC	Radio Resource Control
RRH	Remote Radio Head
RRM	Radio Resource Management
SC-FDMA	Single-carrier Frequency Division Multiple Access
SDMA	Space-Division Multiple-Access
SIC	Successive Interference Cancellation
SINR	Signal-to-Interference-Plus-Noise-Ratio
SNG	Strategic Noncooperative Game
SNR	Signal-to-Noise-Ratio
SOC	Second-Order Cone
SOCP	Second-Order Conic Programming
SON	Self-Organizing Network
TDD	Time Division Multiplexing
WF	Water-Filling
WMMSE	Weighted Minimization of the Mean Squared Error
WSR	Weighted Sum-rate
ZF	Zero-Forcing

Chapter 1
Introduction

1.1 Interference Management in Wireless Cellular Networks

A defining characteristic of a wireless channel is its broadcast nature. In addition, the limited wireless spectrum resource constrains many wireless devices to share the same communication channel, thus inducing mutual inter-user interference. As the interference restricts the reusability of the spectrum resource and the performance among communicating entities, it has always been a critical issue in the deployment of any wireless system. In a multicell environment, the inter-user interference can come from two sources: other devices in the same cell, i.e., intra-cell interference, and co-channel interference from other cells, i.e., inter-cell interference (ICI).

In order to utilize the spectrum resource and control the interference in a better manner, fractional frequency reuse has been widely adopted in early wireless cellular systems, such as Global System for Mobility (GSM). As illustrated in Fig. 1.1 for a fractional frequency reuse system, adjacent cells are guaranteed to operate in different frequencies [18]. On the contrary, cells operating on the same frequency are sufficiently apart such that ICI is kept sufficiently low. Thus, in this conventional multicell system, the ICI is controlled by deploying the frequency reuse pattern and setting maximum power spectral density levels at all the base-stations (BS). As a result, the interference management is usually relegated to a per-cell basis, and the ICI is treated as background noise. It is worth noting that while fractional frequency reuse is adequate in controlling the ICI, each cell is allowed to utilize only parts of the available spectrum. Thus, fractional frequency reuse is inefficient in spectral utilization.

To improve the spectral efficiency, current designs of wireless networks adopt universal frequency reuse where all cells have the potential to use all available radio resources. This implementation is necessary for the current Forth Generation (4G) and future wireless networks to cope with the fast increasing number of wireless mobile-stations (MS) and their demand for higher transmission rates. However, universal frequency reuse comes at the cost of severe ICI, especially at the

D.H.N. Nguyen and T. Le-Ngoc, *Wireless Coordinated Multicell Systems:*
Architectures and Precoding Designs, SpringerBriefs in Computer Science,
DOI 10.1007/978-3-319-06337-9_1, © The Author(s) 2014

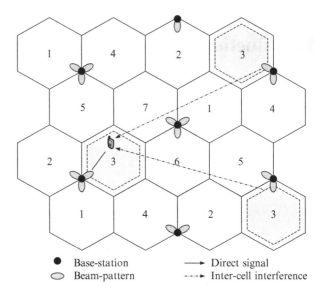

Fig. 1.1 Example of a
fractional frequency reuse
multicell network with a
reuse factor of 7

cell-edge terminals. As depicted in Fig. 1.2, the direct transmission to a particular
MS may be strongly corrupted by the ICI from the adjacent cells. While achieving
higher spectrum efficiency, universal frequency reuse may reduce the sum-rate of
the network if the ICI is not adequately managed.

In recent Third Generation Partnership Project (3GPP) Long-term Evolution
(LTE) Releases, several forms of interference avoidance and coordination techniques were proposed with the main objective to efficiently control the ICI,
especially at the cell-edge MSs [1]. In inter-cell interference coordination (ICIC),
such techniques as time-domain solutions (e.g., subframe alignment) and frequency-
domain methods (e.g., channel orthogonalization) explicitly control how the radio
resources are utilized to moderate the ICI. This approach in dealing with the ICI
might be regarded as *passive* [3]. On the contrary, the emerging wireless communication standard advocates a more *active* treatment of interference through some
forms of interference-aware multicell coordination. In this more advanced coordination technique, namely coordinated multipoint transmission/reception (CoMP),
the inter-cell transmission, instead of being considered as the source of interference,
is taken into account as an extra means to enhance the overall system performance.
The underlying concept of CoMP is quite simple: the coordinated BSs no longer
adjust their parameters (such as precoding, subcarrier assignment, etc.) or decode
independently of each other, but instead coordinate the precoding or decoding
processes on the availability of channel state information (CSI) and the amount of
information signaling over the backhaul links among the BSs [3]. One interesting
aspect of exploiting coordination among the cells is that a fairly small change of
infrastructure is required to implement CoMP. For this reason, it is envisioned that
CoMP will be a key technology of LTE-Advanced [1, 20].

Fig. 1.2 Example of a
multicell network with
universal frequency reuse

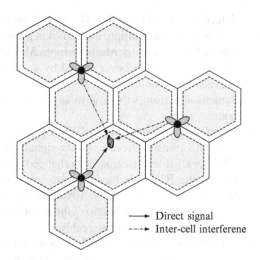

→ Direct signal
---→ Inter-cell interferene

1.2   Coordinated Multipoint Transmission/Reception

The latest 3GPP LTE-Advanced Release considers CoMP as an enabling technology
to improve coverage, throughput and efficiency [1]. Actively dealing with the
ICI, this solution takes advantage of the inter-cell transmissions to enhance the
overall system performance. In the downlink, CoMP coordinates the simultaneous
transmissions from multiple BSs to the MSs. Such coordination is especially helpful
for the cell-edge MSs, whose link conditions are usually unfavorable due to the
long distances from their corresponding BSs, while being more susceptible to
the ICI. In the uplink, CoMP allows for the exploitation of multiple receptions
at multiple BSs to jointly decode the uplink signals from the MSs. CoMP is
especially attractive since it improves the data rate and coverage for cell-edge
users and increases the wireless network's spectral efficiency. It is mentioned in
[1] that CoMP is able to significantly improve the link performance in both uplink
and downlink transmissions. However, this performance enhancement may come
at the cost of excessively high complexity in the joint precoding/decoding process
and the demand for ideal backhaul transmissions and synchronization among the
coordinated BSs.

Following the convention in literature and LTE-Advanced standard, CoMP can
be classified into the following three modes according to the extent of coordination
among the multiple cells [6]

- **Joint Signal Processing (JP)**: Very tight coordination among the cells is
  assumed to perform joint signal processing. User data is exchanged among the
  coordinated BSs such that the multiple BSs can simultaneously transmit/receive
  data signals to/from the MSs within the coordinated area of multiple cells.
  JP is further categorized into Joint Transmission (JT) and Dynamic Point

Selection (DPS) [1,12]. For JT the transmission to a single MS is simultaneously transmitted from multiple transmission points, across cell sites. This mode is also referred to as *Network* Multiple-Input Multiple-Output (MIMO). In DPS, at any one time, the MS is being served by a single transmission point while the data is available at multiple coordinated BSs. However, this single point can change dynamically from subframe to subframe within a set of possible transmission points.

• **Interference Coordination (IC)**: User data is *not* exchanged among the BSs. Each MS transmits/receives *data* signals *only* to/from its (single) serving BS, while *control* information is exchanged among the coordinated BSs to jointly control the ICI. This mode is also mentioned as the coordinated scheduling/ beamforming mode in LTE-Advanced [1]. Hereafter, the "coordination" mode, the "interference coordination" mode, or "coordinated scheduling/beamforming mode" is referred interchangeably.

• **Interference Aware (IA)**: There is no information exchange among the trans-mitting entities. However, the interference is estimated at each receiving entity and fed back to its corresponding transmitter. Acting as a rational entity, each BS selfishly adjusts its transmit/receive strategy according to the knowledge of the interference measured by itself or at its connected MSs. Essentially, the IA mode represents a strategic noncooperative game (SNG) with the BSs being the players. Hereafter, the "competition" mode or the "interference aware" mode is referred interchangeably.

Different CoMP modes impose different requirements on the data and signaling exchanges, as well as the CSI knowledge needed at the coordinated BSs. There is also a trade-off between the performance gain obtained by a CoMP mode and the amount of signaling overhead it placed on the backhaul links. Specifically, the higher extent of coordination among the cells will extract higher performance gains from the multicell network at the cost of higher signaling overhead.

In the JP mode, the antennas from the multiple BSs form a large single antenna array [3,4,7,22]. Data streams intended for all the MSs are jointly processed and can be transmitted from all the antennas. Apparently, while achieving the best performance from the multicell network, such an approach is the most complex CoMP mode (i.e., with the highest level of coordination). In a lesser level of coordination, the IC mode allows a BS to transmit the data only to the MSs within its cell [2,9,17,19]. However, for both JP and IC modes, each coordinated BS may need to know the CSI of all the MSs in the system, even that of unconnected MSs in order to fully control the ICI. In addition, these two modes may require a significant amount of signaling (CSI exchange) to be exchanged among the BSs via an ideal backhaul. On the contrary, under the IA mode, each BS is only required to know the CSI of its connected MSs [5,8,10,11,13–16,21]. Moreover, the IA mode, corresponding to the lowest level of coordination, only demands the signaling on a per-cell basis.

It can be seen that a CoMP system under the JP mode mimics a single large MIMO system with joint transmission/reception processing/antenna selection.

Thus, existing research in single-cell MIMO can be readily adopted to the JP mode. For this reason, this SpringerBrief focuses on the design and signal processing for the CoMP network under the IA and IC modes.

1.3 Organization of This SpringerBrief

This SpringerBrief is to present the architectures of a CoMP system and examine recent advances in precoding designs for such a system. The brief is structured into the following chapters.

Chapter 2 presents an overview on the architecture of CoMP currently deployed in practical networks. We will review the various architectures available for CoMP, with references to the LTE-Advanced standards. The practical implementation and operating challenges of a CoMP system are then discussed in detail.

Chapter 3 presents various state-of-the-art precoding designs in MIMO wireless communications, which serve as the background for the development of precoding designs in CoMP systems in subsequent chapters.

Chapter 4 is concerned with the game theoretical approach in designing the multiuser downlink beamformers in multicell systems. Sharing the same physical resource, the base-station of each cell wishes to minimize its transmit power subject to a set of target signal-to-interference-plus-noise ratios (SINRs) at the multiple users in the cell. In this context, at first, the chapter considers a strategic noncooperative game (SNG) where each base-station greedily determines its optimal downlink beamformer strategy in a distributed manner, without any coordination between the cells. Via the game theory framework, it is shown that this game belongs to the framework of standard functions. The conditions guaranteeing the existence and uniqueness of a Nash Equilibrium (NE) in this competitive design are subsequently examined. In order to improve the efficiency of the NE in the competitive design, this chapter then considers a more cooperative game through a pricing mechanism. The pricing consideration enables a base-station to steer its beamformers in a more cooperative manner, which ultimately limits the interference induced to other cells.

Chapter 5 investigates a multiuser multicell system where block-diagonalization (BD) precoding is utilized on a per-cell basis under two operating modes: IA and IC. In the IA mode, the precoding design in the multicell system can be considered a strategic non-cooperative game (SNG), where each base-station (BS) greedily determines its BD precoding strategy in a distributed manner, based on the knowledge of the inter-cell interference at its connected mobile-stations (MS). Via the game-theory framework, the existence and uniqueness of a Nash equilibrium in this SNG are subsequently studied. In the IC mode, the BD precoders are jointly designed across the multiple BSs to maximize the network weighted sum-rate (WSR). Since this WSR maximization problem is nonconvex, this chapter proposes a distributed algorithm to obtain at least a locally optimal solution. Finally, the analysis of the multicell BD precoding is extended to to the case of BD-Dirty Paper Coding (BD-DPC) precoding.

Chapter 6 is concerned with the maximization of the weighted sum-rate (WSR) in the multicell MIMO multiple access channel (MAC). Considered is the uplink transmission in a multicell network with multiple mobile stations (MS) per cell. Assuming the IC mode, the uplink precoders are jointly optimized at MSs in order to maximize the network WSR. Since this WSR maximization problem is shown to be nonconvex, obtaining its globally optimal solution is rather computationally complex. Thus, the focus of this chapter is on devising a low-complexity algorithm to obtain at least a locally optimal solution. Specifically, by applying successive convex approximation, the nonconvex WSR maximization problem is decomposed into a sequence of multiple MAC sum-rate maximization problems with interference-penalty terms. Our approach will reveal the structure of the optimal uplink precoders as well as the message exchange mechanism to facilitate the distributed implementation of the proposed algorithm. Simulation results then show a significant improvement in the network sum-rate by the proposed algorithm, compared to the case with no interference coordination.

Chapter 7 studies the precoding designs to maximize the weighted sum-rate (WSR) in a multicell MIMO broadcast channel (BC). Considered is the downlink transmission in a coordinated multicell network with multiple mobile stations (MS) per cell. With IC between the multiple cells, the base-station (BS) at each cell only transmits information signals to the MSs within its cell using the dirty paper coding (DPC) technique, while coordinating the inter-cell interference (ICI) induced to other cells. The focus of this chapter is to jointly optimize the encoding covariance matrices across the BSs in order to maximize the network WSR. Since this optimization problem is shown to be nonconvex, obtaining its globally optimal solution is highly complicated. To address this problem, a low-complexity solution approach with distributed implementation is proposed to obtain at least a locally optimal solution. More specifically, by applying a successive convex approximation technique, the original nonconvex problem is decomposed into a sequence of simpler problems, which can be solved optimally and separately at each BS. Simulation results confirm the convergence of the proposed algorithm, as well as their superior performances over schemes with linear precoding or no interference coordination among the BSs.

References

1. 4G Americas: 4G Mobile Broadband Evolution: 3GPP Release 11 & Release 12 and Beyond (2014)
2. Dahrouj, H., Yu, W.: Coordinated beamforming for the multicell multi-antenna wireless system. IEEE Trans. Wireless Commun. **9**(5), 1748–1759 (2010)
3. Gesbert, D., Hanly, S., Huang, H., Shitz, S.S., Simeone, O., Yu, W.: Multi-cell MIMO cooperative networks: a new look at interference. IEEE J. Select. Areas in Commun. **28**(9), 1380–1408 (2010)
4. Karakayali, M., Foschini, G., Valenzuela, R.: Network coordination for spectrally efficient communications in cellular systems. IEEE Wireless Commun. **13**(4), 56–61 (2006)

5. Larsson, E., Jorswieck, E.: Competition versus cooperation on the MISO interference channel. IEEE J. Select. Areas in Commun. **26**(7), 1059–1069 (2008)
6. Marsch, P., Fettweis, G.: Coordinated Multi-point in Mobile Communications: From Theory to Practice. Cambridge University Press, New York: USA (2011)
7. Ng, B.L., Evans, J., Hanly, S., Aktas, D.: Distributed downlink beamforming with cooperative base stations. IEEE Trans. Inform. Theory **54**(12), 5491–5499 (2008)
8. Nguyen, D.H.N., Le-Ngoc, T.: Multiuser downlink beamforming in multicell wireless systems: A game theoretical approach. IEEE Trans. Signal Process. **59**(7), 3326–3338 (2011)
9. Nguyen, D.H.N., Le-Ngoc, T.: Sum-rate maximization in the multicell MIMO broadcast channel with interference coordination. IEEE Trans. Signal Process. **62**(6), 1501–1513 (2014)
10. Nguyen, D.H.N., Le-Ngoc, T.: Sum-rate maximization in the multicell MIMO multiple-access channel with interference coordination. IEEE Trans. Wireless Commun. **13**(1), 36–48 (2014)
11. Pang, J.S., Scutari, G., Facchinei, F., Wang, C.: Distributed power allocation with rate constraints in Gaussian parallel interference channels. IEEE Trans. Inform. Theory **54**(8), 3471–3489 (2008)
12. Sawahashi, M., Kishiyama, Y., Morimoto, A., Nishikawa, D., Tanno, M.: Coordinated multipoint transmission/reception techniques for LTE-Advanced. IEEE Wireless Commun. **17**(3), 26–34 (2010)
13. Scutari, G., Palomar, D.P., Barbarossa, S.: Asynchronous iterative waterfflling for Gaussian frequency-selective interference channels. IEEE Trans. Inform. Theory **54**(7), 2868–2878 (2008)
14. Scutari, G., Palomar, D.P., Barbarossa, S.: Optimal linear precoding strategies for wideband noncooperative systems based on game theory – Part I: Nash Equilibria. IEEE Trans. Signal Process. **56**(3), 1230–1249 (2008)
15. Scutari, G., Palomar, D.P., Barbarossa, S.: Optimal linear precoding strategies for wideband noncooperative systems based on game theory – Part II: Algorithms. IEEE Trans. Signal Process. **56**(3), 1250–1277 (2008)
16. Scutari, G., Palomar, D.P., Barbarossa, S.: The MIMO iterative waterfilling algorithm. IEEE Trans. Signal Process. **57**(5), 1917–1935 (2009)
17. Shi, Q., Razaviyayn, M., Luo, Z.Q., He, C.: An iteratively weighted MMSE approach to distributed sum-utility maximization for MIMO interfering broadcast channel. IEEE Trans. Signal Process. **59**(9), 4331–4340 (2011)
18. Stuber, G.: Principles of Mobile Communication, 3rd Ed. Springer, New York: USA (2011)
19. Venturino, L., Prasad, N., Wang, X.: Coordinated linear beamforming in downlink multi-cell wireless networks. IEEE Trans. Wireless Commun. **9**(4), 1451ï£¡–1461 (2010)
20. Yang, C., Han, S., Hou, X., Molisch, A.: How do we design CoMP to achieve its promised potential? IEEE Wireless Commun. **20**(1), 67–74 (2013)
21. Yu, W., Ginis, G., Cioffi, J.M.: Distributed multiuser power control for digital subscriber lines. IEEE J. Select. Areas in Commun. **20**(5), 1105–1115 (2002)
22. Zhang, H., Dai, H.: Cochannel interference mitigation and cooperative processing in downlink multicell multiuser MIMO networks. EURASIP J. Appl. Signal Process. **2**, 222–235 (2004)

Chapter 2
CoMP: Architectures and Implementations

This chapter presents an overview on the architecture of CoMP currently deployed in practical networks. We will review the various architectures available for CoMP, with references to the LTE-Advanced standard. The practical implementation and operating challenges of a CoMP system are then discussed in detail.

2.1 CoMP Deployment Scenarios in LTE-Advanced

2.1.1 System Requirements Related to CoMP in LTE-Advanced

The 3GPP LTE standard was developed between 2004 and 2009 to provide a high-data-rate, low-latency, and packet-optimized radio-access technology. In 2008, 3GPP started LTE-Advanced standardization as an evolution of Release 8 LTE. Release 10 and 11 LTE-Advanced standards incorporated new features including carrier aggregation, latency reductions, enhanced multi-antenna transmission, small cells, and coordinated multipoint transmission/reception (CoMP) [1–3]. All these new features aim to further enhance the throughput of LTE networks.

The air interface attributes of the LTE and LTE-Advanced systems are summarized in Table 2.1. LTE allows flexible spectrum deployments, where 1.4, 3, 5, 10, 15 and 20 MHz wide cells are standardized. The carrier aggregation feature in LTE-Advanced allows support up to 100 MHz band. LTE system can support the peak download rate up to 300 Mbps with 4×4 MIMO configuration using 20 MHz of spectrum. Since early LTE release does not support MIMO multiplexing in the uplink, the peak upload rate is only 75 Mbps. In LTE-Advanced, the carrier aggregation feature enhances the system throughput to 3 Gbps in the downlink with 8×8 MIMO using 100 MHz band. In LTE-Advanced uplink, MIMO multiplexing is now supported with 4×4 configuration, which boosts the uplink data rate to 1.5 Gbps.

D.H.N. Nguyen and T. Le-Ngoc, *Wireless Coordinated Multicell Systems:* 9
Architectures and Precoding Designs, SpringerBriefs in Computer Science,
DOI 10.1007/978-3-319-06337-9__2, © The Author(s) 2014

Table 2.1 The LTE and LTE-Advanced air interfaces

	Downlink/Uplink	LTE	LTE-Advanced
Spectrum	–	1.4~20 MHz	1.4~100 MHz
Duplexing	–	TDD, FDD, half-duplex FDD	
Peak data rate	Downlink	300 Mbps	3 Gbps
	Uplink	75 Mbps	1.5 Gbps
Multiple access	Downlink	OFDMA	
	Uplink	SC-FDMA	
MIMO	Downlink	2×2, 4×2, 4×4	8×8, 8×4
	Uplink	1×2, 1×4	4×4, 4×8
Modulation	–	QPSK, 16-QAM, 64-QAM	
Channel coding	–	Turbo code	

LTE has adopted orthogonal frequency-division multiple access (OFDMA) in the downlink and single-carrier frequency-division multiple access (SC-FDMA) in the uplink. Due to low peak-to-average power ratio (PAPR) relative to OFDMA [10], SC-FDMA promises lower power consumption and cost at MS and improvement in cell-edge performance, uplink coverage, and capacity. In addition, OFDM-based multiple access techniques in LTE facilitate intracell orthogonality among simultaneous accessing MSs. Thus, the intra-cell interference can be effectively mitigated by good orthogonality of sub-carriers and appropriate physical layer design. On the other hand, there is considerable inter-cell interference (ICI), especially at cell-edge users, due to universal frequency reuse between neighboring cells. To deal with ICI, LTE and LTE-Advanced support various forms of interference avoidance or coordination techniques, including inter-cell interference coordination (ICIC) and coordinated multipoint transmission/reception (CoMP).

In ICIC, frequency resource allocation between neighboring cells are coordinated by explicitly applying restrictions to the radio resource management (RRM) block in time and frequency. This coordinated resource management can be facilitated through fixed, adaptive, or real-time coordination with the help of additional inter-cell signaling. Self-configuration and self-optimization of control parameters of RRM ICIC schemes for uplink and downlink allows proper tuning of ICIC configuration parameters, such as reporting thresholds/periods and resource preference configuration settings, in order to make the ICIC schemes effective with respect to operators' requirements. Enhancements to ICIC, namely eICIC, are being considered in heterogeneous network deployment scenarios with macrocells, picocells, and femtocells [2]. We refer the readers to [5] for a detailed description on ICIC techniques. It is to be noted that the ICIC approach in dealing with the ICI might be regarded as *passive*, compared to CoMP approach.

CoMP is considered by 3GPP as a tool to improve coverage, cell-edge throughput and/or spectral efficiency. A study item was initiated in 3GPP to evaluate this technology in Release 11, followed by a work item for Release 11 which was completed in December 2012. In supporting CoMP, LTE-Advanced Release 11 has provisioned a new Physical Downlink Shared Channel (PDSCH) Transmission Mode, which

includes a common feedback and signaling framework. The common framework allows for multiple non-zero-power Channel-State Information-Reference Symbol (CSI-RS) resources, zero power CSI-RS resources, and interference measurement Channel-State Information-Interference Measurement (CSI-IM) resources to be configured for a MS via Radio Resource Control (RRC) signaling for the measurements of the channel and interference, respectively [3]. The set of CSI-RS resources, which is being used by the MS to measure, includes

- Rank indicator (RI) to indicate the selected number of transmission layers
- Precoding matrix index (PMI) to indicate the selected precoding matrix
- Channel quality indicator (CQI) to indicate the selected modulation scheme.

These indicators compose a CoMP measurement set. In order to avoid large uplink signaling overhead, the maximum size of the CoMP measurement set is three CSI-RS resources. Each MS can process multiple CoMP measurement sets, corresponding to multiple potential CoMP transmission points. Up to 4 CSI feedback processes can be configured for a MS, depending on the MS capability [3]. In addition, due to the mobility of the MS, the CoMP measurement set needs to be reconfigured when necessary to ensure optimal CoMP performance [7].

2.1.2 Deployment Scenarios

CoMP transmissions can be deployed in either inter-site or intra-site as shown in Figs. 2.1 and 2.2. Inter-site CoMP involves the coordination of multiple sites for CoMP transmission. Consequently, the exchange of information involves backhaul links. Thus, inter-site CoMP may put additional overhead upon the backhaul X2 interface connecting the sites. Intra-site CoMP is more advantageous in terms of information exchange between the coordinated cells since backhaul connections between BSs are not involved.

LTE-Advanced support for CoMP was designed such that it can be utilized for different deployment scenarios. Four scenarios were evaluated by 3GPP under the assumption of ideal backhaul characteristics [3, 6, 7]:

- Scenario 1: Homogeneous macro-cellular network with intra-site CoMP
- Scenario 2: Homogeneous macro-cellular network with inter-site CoMP
- Scenario 3: Heterogeneous network with CoMP operation between the macrocell and low power picocells within the macrocell coverage area, where the picocells have different cell identifications (IDs) from the macrocell
- Scenario 4: Heterogeneous network with low power remote radio heads (RRHs) within the macrocell coverage where the transmission/reception points created by the RRHs have the same cell IDs as the macro cell.

Scenarios 1 and 2, depicted in Fig. 2.1 with 3 macrocells per site, are for homogeneous macro-cellular network deployment. In scenario 1, the coordination area is restricted to the cells of a single site, controlled by a BS. Since the extent of

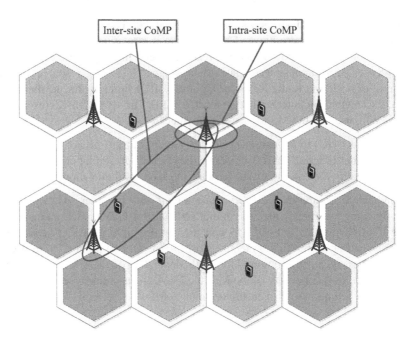

Fig. 2.1 CoMP with inter-site coordination in homogeneous macro-cellular network

coordination is limited to cells of the same site, CoMP scenario 1 is benefitting from
the need for external connections between different sites. Thus, CoMP scenario 1
likely to be the first scenario for practical deployment of CoMP [7]. CoMP scenario
2 is an expanded version of CoMP scenario 1 where the coordination area includes
the cells of different sites. CoMP scenario 2 may be realized by having a single
base-station controller (BSC) to control different sites or by having multiple BSs
at different sites coordinate with each other. Depending on the number of cells
coordinated and the latency of connections between the sites, different levels of
performance gain over CoMP scenario 1 would be possible [7].

Scenarios 3 and 4, specifically tailored for heterogeneous networks, are illustrated in Fig. 2.2. CoMP scenario 3, depicted on the left of Fig. 2.2, is for the
networks where macrocells with high transmission power and picocells with low
transmission power coexist. Each picocell has a different cell ID from the macrocell
and is served by a RRH. Coordination is done between a single macrocell and one or
more picocells located within the macrocell's coverage area. CoMP scenario 3 may
be realized by having a macrocell's BS controlling low-power RRHs of picocells
within the macrocell coverage. The data and control information exchanges between
the BS and the RRHS are transmitted over high capacity and low latency fiber
links. CoMP scenario 4 is slightly different to scenario 3, where the low-power
RRHs share the same physical cell ID as the macrocell. As depicted on the right of
Fig. 2.2, the low power RRHs form a set of distributed antennas of the macrocell.

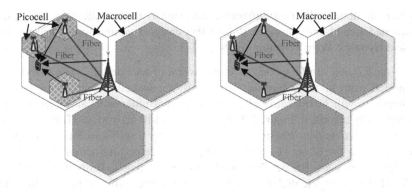

Fig. 2.2 CoMP with intra-site coordination in heterogeneous cellular networks. *Left:* each RRH forms a picocell with different ID. *Right:* each RRH serves as distributed antennas for the macrocell BS

Coordination is done among distributed antennas within a single cell; hence, there is no need for conventional mobility support such as handover procedures among the RRHs. Both CoMP scenarios 3 and 4 are expected to be useful for metropolitan areas where network deployment is dense and RRHs of different transmission power levels coexist [7].

2.2   Classes of CoMP

In the previous section, CoMP is classified based on deployment scenarios. For both homogeneous and heterogeneous networks, the inter- or intra-cell interference is taken into account to enhance the CoMP network's performance. Depending on the geographical separation of the antennas, the coordinated multi-point processing method (for example, coherent or non-coherent), and the coordinated zone definition (for example, cell-centric or user-centric), network MIMO and collaborative MIMO have been proposed for the evolution of LTE [2]. Depending on whether the same data to a UE is shared at different cell sites, CoMP may include single-cell antenna processing with multi-cell coordination, or multi-cell antenna processing. Thus, based on the extent of coordination among the cells, CoMP can be classified into the following modes: Joint Signal Processing (JP), Interference Coordination (IC), and Interference Aware (IA). To this end, each CoMP mode is discussed in detail.

2.2.1   Joint Signal Processing

JP mode corresponds to the highest level of coordination among the cells. In JP mode, user data is exchanged among the coordinated BSs such that the multiple

BSs can simultaneously transmit/receive data signals to/from the MSs within the coordinated area of multiple cells. JP is further categorized into Joint Transmission (JT) and Dynamic Point Selection (DPS) [3, 11].

Joint Transmission: In JT mode, the transmission to a single UE is simultaneously transmitted from multiple transmission points, across cell sites. The multi-point transmissions will be coordinated as a single transmitter with antennas that are geographically separated. Figure 2.3 illustrates a CoMP network in JT mode, where solid arrows denote data transmissions from serving BSs. To facilitate multipoint transmission to the MSs, JT involves cooperative precoding across the multiple cell sites where each BS performs multi-user precoding towards multiple MSs. While JT has the potential for highest performance gain, compared to other coordination mode, it comes at the expense of more stringent requirement on backhaul communications for data and CSI exchange.

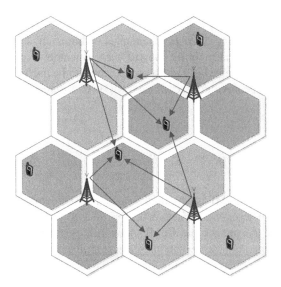

Fig. 2.3 CoMP in the Joint Transmission mode, where *solid arrows* denote data transmissions from serving BSs

Dynamic Point Selection: For DPS, the MS, at any one time, is being served by a single transmission point. However, this single point can dynamically change from time-frame to time-frame within a set of possible transmission points. Figure 2.4 presents an example of CoMP under DPS where solid arrows denote data transmissions from serving BSs and dashed-dotted lines refer to interfering links that could potentially be transmitting data signals in subsequent time-frames. It is to be noted that each MS's data has to be available at all possible transmitting BSs ready for selection. DPS allows CoMP networks to exploit closed-loop macro diversity by selecting the best BS among multiple coordinated sites to serve the MS.

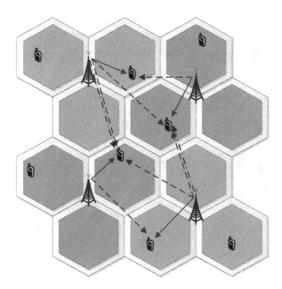

Fig. 2.4 CoMP in the Dynamic Point Selection mode, where *solid arrows* denote data transmissions from serving BSs and *dashed-dotted arrows* refer to interfering links that could potentially be transmitting data signals in subsequent time-frames

2.2.2   Interference Coordination

Unlike the JP mode, the IC mode does not require data sharing between cells. The data for a particular MS is available only at its serving BS and is transmitted from that BS. However, control information, such as precoding decisions and user scheduling are jointly made and exchanged among the coordinated BSs. IC can be implemented via precoding with ICI nulling/mitigation by exploiting the additional degrees of spatial freedom at a cell site. Figure 2.5 illustrates the principles of CoMP in the IC mode, where the coordinated are serving the MSs through data transmission links (solid arrows) while coordinating the ICI to neighboring cells (dashed arrows).

In *coordinated beamforming*, IC allows the BSs to use a coordinated design of the transmit precoder in each cell in order to jointly control and/or mitigate the ICI. In *coordinated user scheduling*, the best serving set of MSs will be selected so that precoders are constructed to reduce the ICI to other neighboring MSs, while increasing the served MSs' signal strength [3]. CoMP under IC modes is applicable for both homogeneous and heterogeneous networks.

To support IC, different CSI processes would be configured for different coordinated transmission points. A given MS's CSI feedback on the CSI process of the serving cell provides the CQI and PMI for the serving cell's transmissions to that UE, while the same UE's PMIs for other transmission points (fed back using other CSI processes) would indicate the precoders to be avoided by the other transmission points, i.e., the precoders that would generate the strongest interference [3].

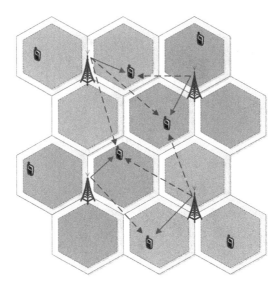

Fig. 2.5 CoMP in the Interference Coordination mode, where *solid arrows* denote data transmissions from serving BSs and *dashed arrows* denote strong ICI from neighboring cells

2.2.3 Interference Aware

Similar to the IC mode, the IA mode requires no data exchange among the transmitting entities, i.e., the data for a particular MS is transmitted from only one connected BS. However, each BS in IA mode has no intention to regulate its induced ICI to neighboring cells. While IA is not mentioned in LTE-Advanced, the analysis of the IA mode provides a base-line for comparison to the achievable performance gain by the IC mode.

In IA, the interference is estimated at each MS and fed back to its serving BS. Acting as a rational entity, each BS then selfishly adjusts its precoding strategy, according to the knowledge of the measured interference. Thus, the IA mode represents a strategic noncooperative game (SNG), where the BSs are the rational players competing for the radio resource. The analysis of the IA mode typically involves the characterization of the Nash equilibrium (NE) of the considered SNG. At an NE, given the precoding strategy from other BSs, a BS does not have the incentive to unilaterally change its precoding strategy. Thus, an NE corresponds to a stable outcome in a non-cooperative game. It is of important concern to predict and even ensure such a state exists and when it is unique.

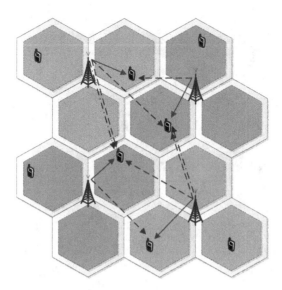

Fig. 2.4 CoMP in the Dynamic Point Selection mode, where *solid arrows* denote data transmissions from serving BSs and *dashed-dotted arrows* refer to interfering links that could potentially be transmitting data signals in subsequent time-frames

2.2.2   Interference Coordination

Unlike the JP mode, the IC mode does not require data sharing between cells. The data for a particular MS is available only at its serving BS and is transmitted from that BS. However, control information, such as precoding decisions and user scheduling are jointly made and exchanged among the coordinated BSs. IC can be implemented via precoding with ICI nulling/mitigation by exploiting the additional degrees of spatial freedom at a cell site. Figure 2.5 illustrates the principles of CoMP in the IC mode, where the coordinated are serving the MSs through data transmission links (solid arrows) while coordinating the ICI to neighboring cells (dashed arrows).

In *coordinated beamforming*, IC allows the BSs to use a coordinated design of the transmit precoder in each cell in order to jointly control and/or mitigate the ICI. In *coordinated user scheduling*, the best serving set of MSs will be selected so that precoders are constructed to reduce the ICI to other neighboring MSs, while increasing the served MSs' signal strength [3]. CoMP under IC modes is applicable for both homogeneous and heterogeneous networks.

To support IC, different CSI processes would be configured for different coordinated transmission points. A given MS's CSI feedback on the CSI process of the serving cell provides the CQI and PMI for the serving cell's transmissions to that UE, while the same UE's PMIs for other transmission points (fed back using other CSI processes) would indicate the precoders to be avoided by the other transmission points, i.e., the precoders that would generate the strongest interference [3].

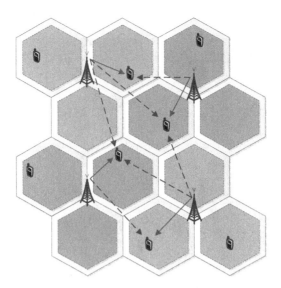

Fig. 2.5 CoMP in the Interference Coordination mode, where *solid arrows* denote data transmissions from serving BSs and *dashed arrows* denote strong ICI from neighboring cells

2.2.3   Interference Aware

Similar to the IC mode, the IA mode requires no data exchange among the transmitting entities, i.e., the data for a particular MS is transmitted from only one connected BS. However, each BS in IA mode has no intention to regulate its induced ICI to neighboring cells. While IA is not mentioned in LTE-Advanced, the analysis of the IA mode provides a base-line for comparison to the achievable performance gain by the IC mode.

In IA, the interference is estimated at each MS and fed back to its serving BS. Acting as a rational entity, each BS then selfishly adjusts its precoding strategy, according to the knowledge of the measured interference. Thus, the IA mode represents a strategic noncooperative game (SNG), where the BSs are the rational players competing for the radio resource. The analysis of the IA mode typically involves the characterization of the Nash equilibrium (NE) of the considered SNG. At an NE, given the precoding strategy from other BSs, a BS does not have the incentive to unilaterally change its precoding strategy. Thus, an NE corresponds to a stable outcome in a non-cooperative game. It is of important concern to predict and even ensure such a state exists and when it is unique.

2.3 Challenges in Implementation and Operation

2.3.1 Challenges in Implementing and Operating CoMP

The deployment of CoMP techniques requires some changes in the implementation of the multicell network. From the network operator's point of view, the change to the current architecture should be limited as much as possible. In this section, we address technical challenges connected to implementing and operating CoMP to realize its full potential.

- *Signaling and Backhaul Transmission*: One of the foremost challenges in operating CoMP is the demand for information and signaling exchange over the backhaul structure. As previously mentioned, CoMP may require instantaneous control signaling, CSI, and even user data (in JP mode) exchanges between coordinated BSs. These requirements stem from a common misconceptions about backhaul links is that the backhaul provides unlimited coordination for the BSs [9]. From practical consideration, both backhaul capacity and latency depends on the existing infrastructure of the network operator. While backhaul links can be upgraded to high-speed optical fiber for higher capacity and lower latency, this creates very high costs to the operators. The latency of the IP-based backhaul X2 interface can cause the delays in the CSI exchange among the coordinated BSs to 10ms or more [12]. From theoretical consideration, the backhaul-constrained coordination raises the tradeoff between the performance gain in a CoMP network and the signaling cost/latency in backhaul transmission.
- *Clustering*: While CoMP under the JP mode can extract the largest performance gains from the multicell network, is requires significant signaling overhead on the air interface and the backhaul. Therefore, only a limited number of BSs within a coordinated area can cooperate to manage the overhead. This raises the question on how to *cluster* the BSs into groups of coordinated cells in order to efficiently exploit the advantages of CoMP at manageable complexity. In general, the BSs can be clustered based on two types of algorithms: static and dynamic. In static clustering, the clusters are kept constant over time based on time-invariant information such as signal propagation properties, geographical positions of BSs and any potential MSs [9]. In dynamic clustering, the clustering strategy is continuously adaptive to changing parameters such as MS locations and radio environment. However, the question remains is on how and where the clustering decisions are made. In addition, any dynamic clustering algorithm is required to fit into the existing architecture of LTE [3]. Interestingly, the self-organizing network (SON) architecture and concept in LTE are applicable to adaptive clustering in CoMP [9].
- *Synchronization*: Another challenge in operating CoMP is the synchronization of cooperated BSs in time and frequency. Synchronization in frequency is necessary to mitigate inter-carrier interference, whereas synchronization in time is to mitigate both inter-symbol and inter-carrier interference. In cellular networks,

synchronization can be obtained by the alignment of the MSs to the time and frequency references from the BSs. However, due to possible intra-site BS coordination, different propagation delays of different terminals may conflict with guard interval, which affects the synchronization [4].

- *Channel Knowledge*: To achieve its full benefit, CoMP transmission requires the knowledge of CSI at the transmitter (CSIT) of both intra-cell and inter-cell channels. CSIT allows the coordinated BSs to not only facilitate multiuser precoding, but also to mitigate ICI. In time-division duplexing (TDD) systems, CSIT can be obtained by exploiting channel reciprocity via uplink training. However, channel reciprocity is no longer true in frequency-division duplexing (FDD) systems due to different uplink and downlink transmission frequencies. Thus, the CSI must be estimated at each MS by downlink training, and then is fed back to the MS's connected BS. The coordinated BSs then rely on the backhaul interface to exchanges the inter-cell channels, as previously mentioned. The CSI feedback and exchange steps may require certain quantizing procedure to the CSI, which then introduces CSIT imperfection. Thus, there are several challenges in obtaining instantaneous CSIT in CoMP networks, including imperfect channel reciprocity in TDD systems, training overhead and feedback, CSI errors due to quantization, CSI exchange overhead on the backhaul interface, and delayed CSIT due to backhaul latency.

- *Efficient and Robust Algorithm*: Many algorithms for CoMP precoding designs come with the requirements for perfect CSIT knowledge across the coordinated cells. As aforementioned, obtaining perfect CSIT is technically challenging in CoMP networks. While precoding designs with imperfect and/or quantized feedbacks in MIMO single-cell systems have attracted a lot of research interests due to their minor performance deficiency, compared to the designs with perfect CSI feedback [8]. Likewise, it is important that the algorithms tailored for CoMP precoding designs must be robust to imperfect CSIT and efficient in terms of performance compared to the case of perfect CSI.

2.3.2 Challenges in Precoding Design for CoMP Systems

Conventionally, most of the related works in precoding designs focus on the single-cell setting where the ICI is simply treated as background noise at the MSs. Thus, the effect of ICI is neglected and the precoding designs are performed independently across the multiple cells. However, with universal frequency reuse, ICI is much more pronounced and should not be ignored. Consequently, existing research in single-cell precoding designs need a rework to take into account the adverse effect of ICI when applied to a multicell system. As discussed in Sect. 1.2, CoMP presents a paradigm shift in precoding designs from the traditional independent per-cell approach to the more coordinated multicell approach. While the potential of performance enhancement of CoMP through the means of precoding is promising, several technical challenges directly related CoMP precoding designs are as follows.

First, it is well known that the performance of a precoding design is largely dependent on the knowledge of CSIT. Thus, it is imperative that the transmitter, acting as a central unit, gathers the CSI to all of its connected users. However, due to the large-scale and distributed nature of the multicell system, it is much more difficult for a central unit to collect the CSI from all coordinated BSs to all MSs. Hence, instead of using a central unit in a CoMP system, the precoders should be devised in a fully distributed manner across the coordinated BSs. Essentially, each coordinated BS should design the precoders for its connected MSs using only local CSI.

Second, due to the distributed implementation in the precoding designs, the coordinated BSs may need to exchange signaling and control information. However, there are restrictions on the sharing of information among the coordinated BSs due to the limitations of the backhaul links. Thus, it is important to define and quantize the amount of message exchange, including the inter-cell messages among coordinated BSs and the intra-cell messages between a BS and its connected MS.

Third, while many precoding design problems are convex under the single-cell setting, they are nonconvex under the CoMP setting. These problems, typically difficult and computationally complex to optimally solve, pose a requirement for efficient suboptimal algorithms. Coupled with the distributed implementation requirement, it is important that certain optimization steps in the algorithms can be assigned to and executed at each BS and MS with local information. In addition, the algorithms should provide a good trade-off between the achievable performance and the computational complexity. Whether or not CoMP precoders can be efficiently designed depends on how the aforementioned technical challenges are addressed.

References

1. 3GPP TR 36.814: Further enhancements for U-TRAN; Physical Layer Aspects (2009)
2. 4G Americas: 4G Mobile Broadband Evolution: 3GPP Release 10 & Release 11 and Beyond - HSPA+, SAE/LTE and LTE-Advanced (2012)
3. 4G Americas: 4G Mobile Broadband Evolution: 3GPP Release 11 & Release 12 and Beyond (2014)
4. Irmer, R., Droste, H., Marsch, P., Grieger, M., Fettweis, G., Brueck, S., Mayer, H.P., Thiele, L., Jungnickel, V.: Coordinated multipoint: Concepts, performance, and field trial results. IEEE Commun. Magazine 49(2), 102–111 (2011)
5. Kosta, C., Hunt, B., Quddus, A., Tafazolli, R.: On interference avoidance through inter-cell interference coordination (ICIC) based on OFDMA mobile systems. IEEE Commun. Surveys & Tutorials 15(3), 973–995 (2013)
6. Lee, D., Seo, H., Clerckx, B., Hardouin, E., Mazzarese, D., Nagata, S., Sayana, K.: Coordinated multipoint transmission and reception in LTE-Advanced: Deployment scenarios and operational challenges. IEEE Commun. Magazine 50(2), 148–155 (2012)
7. Lee, J., Kim, Y., Lee, H., Ng, B.L., Mazzarese, D., Liu, J., Xiao, W., Zhou, Y.: Coordinated multipoint transmission and reception in LTE-Advanced systems. IEEE Commun. Magazine 50(11), 44–50 (2012)

8. Love, D.J., R. W. Heath, J., Lau, V.K.N., Gesbert, D., Rao, B.D., Andrews, M.: An overview of limited feedback in wireless communication systems. IEEE J. Select. Areas in Commun. **26**(8), 1341–1365 (2008)
9. Marsch, P., Fettweis, G.: Coordinated Multi-point in Mobile Communications: From Theory to Practice. Cambridge University Press, New York: USA (2011)
10. Myung, H.G., Lim, J., Goodman, D.: Single carrier FDMA for uplink wireless transmission. IEEE Veh. Tech. Magazine **1**(3), 30–38 (2006)
11. Sawahashi, M., Kishiyama, Y., Morimoto, A., Nishikawa, D., Tanno, M.: Coordinated multipoint transmission/reception techniques for LTE-Advanced. IEEE Wireless Commun. **17**(3), 26–34 (2010)
12. Yang, C., Han, S., Hou, X., Molisch, A.: How do we design CoMP to achieve its promised potential? IEEE Wireless Commun. **20**(1), 67–74 (2013)

Chapter 3
Multiuser Precoding Designs

This chapter presents various state-of-the-art precoding designs in MIMO communications, which serve as the background for the development of precoding designs in CoMP systems in subsequent chapters. The first part of this chapter discusses precoding techniques in multiuser single-cell systems under two design objectives: power minimization and sum-rate maximization. The second part of this chapter then presents some recent advances in multicell precoding designs.

3.1 Precoding Designs in a MIMO Single-Cell System

3.1.1 Downlink Beamforming for Power Minimization

Consider a single cell system comprising of K single-antenna MSs, concurrently served by an M-antenna BS, as depicted in Fig. 3.1. Using linear precoding, the BS multiplexes several data streams into its transmitted signal $\mathbf{x}$ as $\mathbf{x} = \sum_{i=1}^{K} u_i \mathbf{w}_i$, where u_i is the signal symbol intended for user-i with unit energy, i.e., $\mathbb{E}[|u_i|] = 1$, and $\mathbf{w}_i$ is an $M \times 1$ beamformer vector designed for user-i. In the downlink transmission, the received signal at user-i can be modeled as

$$y_i = \mathbf{h}_i^H \mathbf{x} + z_i = \mathbf{h}_i^H \sum_{j=1}^{K} u_j \mathbf{w}_j + z_i, \qquad (3.1)$$

where $\mathbf{h}_i^* \in \mathbb{C}^{M \times 1}$ is the channel vector to user-i and $z_i \sim \mathcal{CN}(0, \sigma^2)$. It is easy to verify that the signal-to-interference-plus-noise ratio (SINR) is given by

$$\mathrm{SINR}_i = \frac{\left| \mathbf{h}_i^H \mathbf{w}_i \right|^2}{\sum_{j \neq i}^{K} \left| \mathbf{h}_i^H \mathbf{w}_j \right|^2 + \sigma^2}, \qquad (3.2)$$

where $\sum_{j \neq i}^{K} \left| \mathbf{h}_i^H \mathbf{w}_j \right|^2$ is the inter-user interference at user-i.

D.H.N. Nguyen and T. Le-Ngoc, *Wireless Coordinated Multicell Systems: Architectures and Precoding Designs*, SpringerBriefs in Computer Science, DOI 10.1007/978-3-319-06337-9_3, © The Author(s) 2014

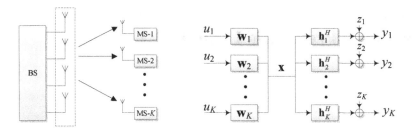

Fig. 3.1 Example of a multiuser MIMO system with downlink beamforming

Note that the performance of a system is usually quantified by the quality of service (QoS) at each MS and the power usage at the BS. In this case, the QoS is defined as the SINR, which is directly related to the achievable rate at the MS. One of the more popular design criteria in multiuser beamforming is that the BS attempts to minimize its transmit power subject to the QoS constraints at MSs. This power minimization problem is stated as

$$\underset{\mathbf{w}_1,\ldots,\mathbf{w}_K}{\text{minimize}} \quad \sum_{i=1}^{K} \|\mathbf{w}_i\|^2 \tag{3.3}$$

$$\text{subject to} \quad \frac{\left|\mathbf{h}_i^H \mathbf{w}_i\right|^2}{\sum_{j\neq i}^{K} \left|\mathbf{h}_i^H \mathbf{w}_j\right|^2 + \sigma^2} \geq \gamma_i,$$

where γ_i is the target SINR at user-i.

This optimization problem was initially considered to be nonconvex [26, 27]. Nonetheless, the problem can be optimally solved by various approaches. Uplink-downlink duality was exploited in [26,27,29,44], where the downlink problem under individual SINR constraints can be solved via the equivalent uplink problem, which is convex and much easier to solve. In another approach in [1], this nonconvex problem was relaxed into a convex semi-definite program (SDP). In more recent works [18, 46, 52], the authors formulated the downlink problem directly as a convex second-order conic program (SOCP). A simple and fast fixed-point iterative algorithm was also proposed to find the optimal downlink beamformers.

3.1.2 Uplink Precoding for Multiple-Access Channel

This section provides a brief review on MIMO multiple-access channel (MAC) in a single-cell multiuser system. The sum-rate maximization in the MAC and optimal designs of the uplink precoding for MAC will be sequentially presented. Consider a system with K MSs, each equipped with N transmit antennas, concurrently transmitting to a BS equipped with M receive antennas, as illustrated in Fig. 3.2.

The uplink transmission can be modeled as

$$y = \sum_{i=1}^{K} \mathbf{H}_i \mathbf{x}_i + \mathbf{z}, \qquad (3.4)$$

where $\mathbf{x}_i \in \mathbb{C}^N$ is the transmitted vector signal from user-i, $\mathbf{y} \in \mathbb{C}^M$ is the received vector signal at the BS, $\mathbf{H}_i \in \mathbb{C}^{M \times N}$ represents the channel matrix from user-i to the BS, and $\mathbf{z}$ is the AWGN at the BS with zero mean and covariance matrix $\mathbf{Z}$. We denote $\mathbf{X}_i = \mathbb{E}[\mathbf{x}_i \mathbf{x}_i^H]$ as the transmit covariance matrix of user-i.

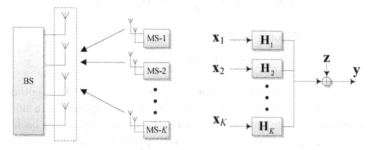

Fig. 3.2 Example of a MIMO multiple-access channel

In the MAC, the capacity-achieving decoder is the successive interference cancellation (SIC) decoder [7,53], where the BS decodes the signals from the multiple users in sequence. After the BS decodes one user's signal, it then suppresses this user's signal from its received signal such that this user's signal no longer interferes the users being decoded later. For example, assume that the decoding order is from user-1 to user-K. Using SIC, after decoding the signal from user-1, the BS suppresses the signal from user-1 from the received signal before processing the signal from user-2. This process continues such that user-i only experiences the interferences from user-$(i + 1)$ to user-K. While treating these interferences as noises, the achievable rate of user-i is [41]

$$R_i^{\text{MAC}} = \log \left| \mathbf{I} + \mathbf{H}_i^H \left(\mathbf{Z} + \sum_{j>i}^{K} \mathbf{H}_j \mathbf{X}_j \mathbf{H}_j^H \right)^{-1} \mathbf{H}_i \mathbf{X}_i \right|$$

$$= \log \frac{\left| \mathbf{Z} + \sum_{j=i}^{K} \mathbf{H}_j \mathbf{X}_j \mathbf{H}_j^H \right|}{\left| \mathbf{Z} + \sum_{j>i}^{K} \mathbf{H}_j \mathbf{X}_j \mathbf{H}_j^H \right|}. \qquad (3.5)$$

The sum-rate in the MAC $\sum_{i=1}^{K} R_i^{\text{MAC}}$ can be simplified as

$$\sum_{i=1}^{K} R_i^{\text{MAC}} = \sum_{i=1}^{K} \log \frac{\left| \mathbf{Z} + \sum_{j=i}^{K} \mathbf{H}_j \mathbf{X}_j \mathbf{H}_j^H \right|}{\left| \mathbf{Z} + \sum_{j>i}^{K} \mathbf{H}_j \mathbf{X}_j \mathbf{H}_j^H \right|} = \log \left| \mathbf{Z} + \sum_{i=1}^{K} \mathbf{H}_i \mathbf{X}_i \mathbf{H}_i^H \right| - \log |\mathbf{Z}|.$$

(3.6)

The goal is to maximize this sum-rate. That is,

$$\underset{\mathbf{X}_1,\dots,\mathbf{X}_K}{\text{maximize}} \ \log \left| \mathbf{Z} + \sum_{i=1}^{K} \mathbf{H}_i \mathbf{X}_i \mathbf{H}_i^H \right| - \log |\mathbf{Z}| \qquad (3.7)$$

$$\text{subject to} \ \ \text{Tr}\{\mathbf{X}_i\} \leq P_i, \ \forall i$$

$$\mathbf{X}_i \succeq \mathbf{0}, \ \forall i,$$

where P_i is maximum allowable transmit power at user-i. This optimization problem is a convex problem, which allow for efficient algorithms in finding its optimal solution. In addition, as the constraints are decoupled for each of the variables $\mathbf{X}_1,\dots,\mathbf{X}_K$, the optimization can be performed sequentially for each variable. In particular, while treating $\mathbf{Z}_i = \mathbf{Z} + \sum_{j \neq i}^{K} \mathbf{H}_j \mathbf{X}_j \mathbf{H}_j^H$ as noise, one may perform the optimization

$$\underset{\mathbf{X}_i}{\text{maximize}} \ \log \left| \mathbf{Z}_i + \mathbf{H}_i \mathbf{X}_i \mathbf{H}_i^H \right| \qquad (3.8)$$

$$\text{subject to} \ \ \text{Tr}\{\mathbf{X}_i\} \leq P_i$$

$$\mathbf{X}_i \succeq \mathbf{0}.$$

This optimization problem can be solved by applying the water-filling (WF) process as follows. First, we perform the eigen-decomposition

$$\mathbf{H}_i^H \mathbf{Z}_i^{-1} \mathbf{H}_i = \mathbf{U}_i \mathbf{D}_i \mathbf{U}_i^H, \qquad (3.9)$$

where $\mathbf{U}_i$ is a semi-unitary matrix and $\mathbf{D}_i$ is diagonal matrix with non-negative eigenvalues. Then, the optimal solution $\mathbf{X}_i^\star$ is obtained from the well-known WF solution [53]

$$\mathbf{X}_i^\star = \mathbf{U}_i \left[\mu_i \mathbf{I} - \mathbf{D}_i^{-1} \right]^+ \mathbf{U}_i^H, \qquad (3.10)$$

where μ_i is the water-level, chosen to meet the power constraint $\text{Tr} \left\{ \left[\mu_i \mathbf{I} - \mathbf{D}_i^{-1} \right]^+ \right\} = P_i$. The solution to the MAC sum-rate maximization is then obtained by iteratively performing the above WF process for each of the variables $X_1, \ldots, X_K$ until we reach the converging state where the MAC sum-rate can no longer be improved.

3.1.3 Downlink Precoding for Broadcast Channel

This section reviews the multiuser Gaussian vector broadcast channel (BC) in a single-cell system. Consider the same system model as in Sect. 3.1.2, as depicted in Fig. 3.3, but the transmission is for the downlink instead. Denote $\mathbf{x} = \sum_{i=1}^{K} \mathbf{x}_i$ as the transmitted signal from the BS, where $\mathbf{x}_i$ represents the transmitted signal for user-i. Further denote $\mathbf{Q}_i = \mathbb{E}[\mathbf{x}_i \mathbf{x}_i^H]$ as the transmit covariance matrix for user-i. The received signal at the ith user is given by

$$\mathbf{y}_i = \mathbf{H}_i^H \mathbf{x} + \mathbf{z}_i = \mathbf{H}_i^H \sum_{j=1}^{K} \mathbf{x}_j + \mathbf{z}_i, \tag{3.11}$$

where $\mathbf{z}_i$ is the additive Gaussian noise with the covariance $\mathbf{Z}_i$ and $\mathbf{H}_i^H \in \mathbb{C}^{N \times M}$ is the channel from the M-antenna BS to the N-antenna MS-i.

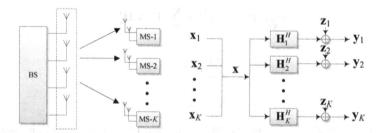

Fig. 3.3 Example of a MIMO broadcast channel

It is well known that the capacity-achieving coding scheme in the BC is "dirty-paper coding" (DPC) [3, 6, 50]. The fundamental idea of DPC is that when a transmitter knows the interference in a channel in advance, it can design a code to compensate for the interference such that the capacity of the channel is the same as if there is no interference. Assuming the encoding order from user-K to user-1, DPC is utilized such that the intended codeword for user-i does not see the intra-cell interference from user-$(i+1)$ to user-K. Thus, the achievable rate at user-i is given by

$$R_i^{\text{BC}} = \log \left| \mathbf{I} + \mathbf{H}_i \left(\mathbf{Z}_i + \sum_{j=1}^{i-1} \mathbf{H}_i^H \mathbf{Q}_j \mathbf{H}_i \right)^{-1} \mathbf{H}_i^H \mathbf{Q}_i \right|$$

$$= \log \frac{\left| \mathbf{Z}_i + \sum_{j=1}^{i} \mathbf{H}_i^H \mathbf{Q}_j \mathbf{H}_i \right|}{\left| \mathbf{Z}_i + \sum_{j=1}^{i-1} \mathbf{H}_i^H \mathbf{Q}_j \mathbf{H}_i \right|}. \tag{3.12}$$

When the objective is to maximize the sum-rate in the BC with DPC, the optimization is given by

$$\underset{\mathbf{Q}_1,\dots,\mathbf{Q}_K}{\text{maximize}} \ \sum_{i=1}^{K} \log \frac{\left| \mathbf{Z}_i + \sum_{j=1}^{i} \mathbf{H}_i^H \mathbf{Q}_j \mathbf{H}_i \right|}{\left| \mathbf{Z}_i + \sum_{j=1}^{i-1} \mathbf{H}_i^H \mathbf{Q}_j \mathbf{H}_i \right|} \tag{3.13}$$

$$\text{subject to} \ \sum_{i=1}^{K} \text{Tr}\{\mathbf{Q}_i\} \leq P$$

$$\mathbf{Q}_i \succeq \mathbf{0}, \ \forall i,$$

where P is total allowable transmit power at the BS. Unlike the MAC problem (3.7), the optimization problem for the BC channel is nonconvex due to the inter-user interference components in the objective function. Fortunately, via the so-called BC-MAC duality [43, 45], the BC problem (3.13) can be transformed into an equivalent MAC problem

$$\underset{\mathbf{X}_1,\dots,\mathbf{X}_K}{\text{maximize}} \ \log \left| \mathbf{I} + \sum_{i=1}^{K} \tilde{\mathbf{H}}_i \mathbf{X}_i \tilde{\mathbf{H}}_i^H \right| \tag{3.14}$$

$$\text{subject to} \ \sum_{i=1}^{K} \text{Tr}\{\mathbf{X}_i\} \leq P$$

$$\mathbf{X}_i \succeq \mathbf{0}, \ \forall i,$$

where $\tilde{\mathbf{H}}_i = \mathbf{H}_i \mathbf{Z}_i^{-1/2}$. Unlike the MAC problem (3.7), the power constraint in problem (3.14) is now a single sum-power constraint on all covariance matrix variables $\mathbf{X}_i$'s. Since the transformed problem (3.14) is now convex, it can be optimally solved by efficient convex optimization methods, including the block WF algorithm [14] and the dual decomposition method [49]. The optimal solution $\mathbf{Q}_i^\star$ of the original BC problem (3.13) then can be found from the optimal solution $\mathbf{X}_i^\star$ of the equivalent MAC problem through the so-called MAC-BC transformation [43].

It is to be noted that DPC can only serves as a theoretical benchmark due to its high complexity implementation that involves random nonlinear coding. Linear transmission techniques are an attractive alternative because of their simplicity. With linear precoding, the achievable rate at MS-i is given by

$$R_i = \log \left| \mathbf{I} + \mathbf{H}_i \left(\mathbf{Z}_i + \sum_{j \neq i}^{K} \mathbf{H}_i^H \mathbf{Q}_j \mathbf{H}_i \right)^{-1} \mathbf{H}_i^H \mathbf{Q}_i \right|$$

$$= \log \frac{\left| \mathbf{Z}_i + \sum_{j=1}^{K} \mathbf{H}_i^H \mathbf{Q}_j \mathbf{H}_i \right|}{\left| \mathbf{Z}_i + \sum_{j \neq i}^{K} \mathbf{H}_i^H \mathbf{Q}_j \mathbf{H}_i \right|}, \tag{3.15}$$

where the inter-user interference is treated as background noise. When the goal is to maximize the sum-rate in the BC, the problem is to maximize $\sum_{i=1}^{K} R_i$. That is,

$$\underset{\mathbf{Q}_1,\ldots,\mathbf{Q}_K}{\text{maximize}} \sum_{i=1}^{K} \log \frac{\left| \mathbf{Z}_i + \sum_{j=1}^{K} \mathbf{H}_i^H \mathbf{Q}_j \mathbf{H}_i \right|}{\left| \mathbf{Z}_i + \sum_{j \neq i}^{K} \mathbf{H}_i^H \mathbf{Q}_j \mathbf{H}_i \right|} \tag{3.16}$$

$$\text{subject to} \sum_{i=1}^{K} \text{Tr}\{\mathbf{Q}_i\} \leq P$$

$$\mathbf{Q}_i \succeq \mathbf{0}, \ \forall i.$$

Like the sum-rate maximization problem with DPC in (3.13), the above optimization is nonconvex. However, it is no longer possible to transform problem (3.16) into an equivalent convex problem in the MAC as in (3.14). Thus, it is generally difficult to obtain its globally optimal solution. Recent works in [5, 38] proposed iterative algorithms that convert this nonconvex problem into a sequence of convex mean squared error (MSE) minimization problems. These algorithms are shown to converge monotonically to a local optimal solution. However, the drawbacks of these algorithms are their high computational costs and intractable solutions. Alternately, one may apply the block diagonalization (BD) precoding [4, 24, 40, 47] where the inter-user interference is completely eliminated, i.e., $\mathbf{H}_i^H \mathbf{Q}_j \mathbf{H}_i = \mathbf{0}$, if $j \neq i$. In this case, the precoding matrix for a particular user, say user-i, $\mathbf{Q}_i$ is limited to be in the null space created by $[\mathbf{H}_1, \ldots, \mathbf{H}_{i-1}, \mathbf{H}_{i+1}, \ldots, \mathbf{H}_K]$. As a result, the objective function under BD constraints becomes convex and the optimal solution can easily be obtained in a closed-form WF solution [40]. It is to be noted that BD precoding is only restricted to the system where the number of transmit antennas at the BS is no smaller than the total number of receive antennas at all the MSs. Nonetheless, even with the BD constraints, the performance gap between BD precoding and the benchmark DPC is negligible at high SNR [40].

3.2 Precoding Designs for MIMO Multicell Systems with CoMP

This section examines some recent advances in precoding designs for MIMO multicell networks employing CoMP under the IA and IC mode. Note that these precoding designs share many similar properties with the precoding designs in single-cell networks. The difference here is the consideration of individual power constraint at each coordinated BS and the inherent ICI among the cells. With per-base-station power constraints, this section reviews CoMP precoding designs under the following two design criteria: (a) minimizing the transmit power at the BSs with a set of target rates at the MSs and (b) maximizing the sum-rate at the connected MSs.

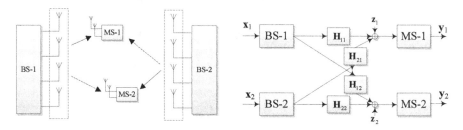

Fig. 3.4 Example of a multicell system with 2 base-stations and 2 mobile-stations

3.2.1   Interference Aware

The study of precoding design and power control in a mutual interference network using game theory has recently attracted considerable research attention. By considering the interference network as a SNG, each player greedily adapts its strategy to maximize its own utility, given the strategies from other players [10,12,13,16,17,25, 28,30,32–34,39,51].[1] In general, these works focus on studying the existence and uniqueness of the stable operating point of the system, i.e., the NE. The uplink power control problem in a single-cell code-division multiple-access (CDMA) data system with multiple competing users was studied in [13,28], where the utility function was defined as the ratio of throughput to transmit power. A pricing mechanism was investigated in [28] to obtain a more efficient solution of the power control game. For an orthogonal frequency division multiplexing (OFDM) system over a shared band, the work in [51] has inspired various works on the iterative water-filling (IWF) algorithm, such as [12,17,30,32,33,39] with sum-rate as the utility function, or [25] with transmit power as the utility function.

In multiple antenna systems, the work in [16] considered a multicell system, where each cell consisted of *one* multiple-antenna BS and *one* single antenna MS. The objective of [16] was to study the precoding beamforming vector at the two BSs in competitive and cooperative manners. In the multiuser MIMO channels, the work in [34] studied the competitive precoding design, where each player wished to maximize its mutual information. It is to be noted that these works only considered

---

[1]Hereafter, a cell or a base-station is referred to as a player interchangeably, whereas a MS is referred to as a user.

the system where each transmitter, i.e, BS, communicates with only *one* receiver, i.e., MS. We now discuss in detail the representative works that are closely related to the multicell network with CoMP.

Consider a MIMO multicell system with Q BSs and Q MSs, as depicted in Fig. 3.4. In each cell, the BS is sending information only to its connected MS. Let $\Omega = \{1, \ldots, Q\}$ denote the set of cells (players). The transmission over the q-th MIMO channel with M_q transmit and N_q receive dimensions can be described by the baseband signal model

$$\mathbf{y}_q = \mathbf{H}_{qq}\mathbf{x}_q + \sum_{r \neq q}^{Q} \mathbf{H}_{rq}\mathbf{x}_r + \mathbf{z}_q, \tag{3.17}$$

where $\mathbf{x}_q \in \mathbb{C}^{M_q \times 1}$ is the transmitted signal vector at BS-q and $\mathbf{y}_q \in \mathbb{C}^{N_q \times 1}$ is the received signal vectors at MS-q, and $\mathbf{z}_q$ is the AWGN with a covariance matrix $\mathbf{Z}_q$. The channel matrix from the BS-r's transmitter to MS-q is represented by $\mathbf{H}_{rq} \in \mathbb{C}^{N_r \times M_q}$. Likewise, $\sum_{r \neq q}^{Q} \mathbf{H}_{rq}\mathbf{x}_r$ is the interference induced by other MIMO links at MS-q.

Denote the precoding strategy at BS-q as $\mathbf{Q}_q = \mathbb{E}\left[\mathbf{x}_q\mathbf{x}_q^H\right]$ and the precoding strategy profile at all BSs except BS-q as $\mathbf{Q}_{-q}$. While treating the ICI from other links as additive Gaussian noise, the achievable rate at the link-q between BS-q and MS-q is given by [41]

$$R_q(\mathbf{Q}_q, \mathbf{Q}_{-q}) = \log\left|\mathbf{I} + \mathbf{H}_{qq}^H\mathbf{R}_q^{-1}(\mathbf{Q}_{-q})\mathbf{H}_{qq}\mathbf{Q}_q\right|, \tag{3.18}$$

where $\mathbf{R}_q(\mathbf{Q}_{-q}) = \mathbf{Z}_q + \sum_{r \neq q}^{Q} \mathbf{H}_{rq}\mathbf{Q}_r\mathbf{H}_{rq}^H$ is the interference plus noise covariance matrix at MS-q.

Depending on the strategies of the other players, which are reflected in $\mathbf{R}_q(\mathbf{Q}_q, \mathbf{Q}_{-q})$, player-$q$ may want to maximize its achievable rate subject to a constraint on its transmit power. In that case, the utility of player-q is defined as $u_q = R_q(\mathbf{Q}_q, \mathbf{Q}_{-q})$, whereas the set of admissible strategies is defined as

$$\mathscr{P}_q = \left\{\mathbf{Q}_q \in \mathbb{C}^{M_q \times M_q} : \mathbf{Q}_q \succeq 0, \operatorname{Tr}\{\mathbf{Q}_q\} \leq P_q^{\max}\right\}. \tag{3.19}$$

Naturally, the players in Ω forms a SNG with the utility u_q and the set of admissible strategies $\mathscr{P}_q$ for player-q. Mathematically, the SNG can be defined as

$$\mathscr{G}_R = \left(\Omega, \{\mathscr{P}_q\}_{q \in \Omega}, \{R_q(\mathbf{Q}_q, \mathbf{Q}_{-q})\}_{q \in \Omega}\right). \tag{3.20}$$

On the contrary, if each player attempts to minimize its transmit power subject to a constraint on its achievable rate, the utility function is defined as $u_q = -\operatorname{Tr}\{\mathbf{Q}_q\}$, whereas the set of admissible strategies is defined as

$$\mathscr{R}_q(\mathbf{Q}_{-q}) = \left\{ \mathbf{Q}_q \in \mathbb{C}^{M_q \times M_q} : \mathbf{Q}_q \succeq \mathbf{0}, R_q(\mathbf{Q}_q, \mathbf{Q}_{-q}) \geq R_q^{\min} \right\}. \qquad (3.21)$$

Thus, the corresponding SNG is given by

$$\mathscr{G}_P = \left(\Omega, \ \{\mathscr{R}_q(\mathbf{Q}_{-q})\}_{q \in \Omega}, \ \{-\mathrm{Tr}\{\mathbf{Q}_q\}\}_{q \in \Omega} \right). \qquad (3.22)$$

For the case of multiple input single output (MISO) channels, i.e., $N_q = 1, \forall q$, the NE characterizations of games $\mathscr{G}_R$ and $\mathscr{G}_P$ are quite straightforward. It is maintained in [16] that the best response (BR) beamforming strategy at each player is the maximal ratio transmitting (MRT) beamformer, i.e., $\mathbf{x}_q = \mathbf{H}_q^H / \|\mathbf{H}_q\|$. This is due to the fact that an MRT beamformer maximizes the SINR at its corresponding receiver. Thus, each player only concerns with its transmit power adjustment to maximize its utility.

Let P_q be the transmit power at BS-q. In game $\mathscr{G}_R$, it is straightforward to see that each player transmits at its maximum power at the NE [16], i.e., $P_q = P_q^{\max}$. Thus the NE of game $\mathscr{G}_R$ is always existent and unique. In game $\mathscr{G}_P$, player-q performs its BR strategy at a given time slot, say $t + 1$, by adjusting its transmit power P_q as

$$P_q[t + 1] = \frac{\mathrm{e}^{R_q^{\min}} - 1}{\mathrm{e}^{R_q[t]} - 1} P_q[t], \quad q \in \Omega, \qquad (3.23)$$

where $\mathrm{e}^{R_q[t]} - 1$ is the measured SINR at MS-q at time t. Note that the above power update is well-known as the optimal power control mechanism for CDMA-based wireless networks [11]. This iteration always converges to a unique fixed-point, which is the NE of game $\mathscr{G}_P$, if such a NE exists [48]. The condition for the existence and uniqueness of game $\mathscr{G}_P$ will be presented later in Chap. 4.

For the case of multiple receive antennas, i.e., $N_q > 1$, the analyses of games $\mathscr{G}_R$ and $\mathscr{G}_P$ are much more difficult. Nonetheless, various works in literature have addressed the NE analysis in these games [25, 31, 34]. For both games, the BR strategy by each player has the WF structure. Specifically, given the eigen-decomposition $\mathbf{H}_{qq}^H \mathbf{R}_q^{-1} \mathbf{H}_{qq} = \mathbf{U}_q \mathbf{D}_q \mathbf{U}_q^H$, the BR strategy at player-q is

$$\mathbf{Q}_q^\star = \mathbf{WF}_q(\mathbf{Q}_{-q}) = \mathbf{U}_q \left[\mu_q \mathbf{I} - \mathbf{D}_q^{-1} \right]^+ \mathbf{U}_q^H, \qquad (3.24)$$

where μ_q is the water-level, chosen either to meet the power constraint in game $\mathscr{G}_R$ or to meet the rate constraint in game $\mathscr{G}_P$. Each player plays the WF strategy until the game converges. For this reason, these games are often referred to as IWF games in literature. The NE of this noncooperative game, which is the intersection between the BR strategies of each user, can be restated as

$$\mathbf{Q}_q^\star = \mathbf{WF}_q(\mathbf{Q}_{-q}^\star), \quad \forall q \in \Omega. \qquad (3.25)$$

Using this WF structure, the sufficient conditions on the existence and uniqueness of these IWF games are presented in [31, 34]. A key observation from those conditions reveals that the game's NE is always existent and unique when the ICI is sufficiently small.

3.2.2 Interference Coordination

In the previous section, a fully decentralized approach to the CoMP under the IA mode was examined using the game theory framework and the NE of the system was characterized. However, it is well-known that the NE need not be Pareto-efficient [9]. Via the interference coordination among the BSs, significant power reduction or rate enhancement can be obtained by jointly designing all the precoders across the coordinated BSs at the same time. Pareto-efficient precoding designs have been recently proposed in [2, 16] for the multicell system with interference coordination.

When the design setting is to jointly maximize the network sum-rate, the joint optimization for CoMP under the IC mode can be stated as

$$\underset{\mathbf{Q}_1,\ldots,\mathbf{Q}_K}{\text{maximize}} \sum_{q=1}^{Q} \omega_q \left| \mathbf{I} + \mathbf{H}_{qq}^H \mathbf{R}_q^{-1}(\mathbf{Q}_{-q})\mathbf{H}_{qq}\mathbf{Q}_q \right| \tag{3.26}$$

$$\text{subject to } \text{Tr}\{\mathbf{Q}_q\} \leq P_q^{\max}, \; \forall q$$

$$\mathbf{Q}_q \succeq \mathbf{0}, \; \forall q, \tag{3.27}$$

where ω_q is the non-negative weight for the link-q. For a given set of weights $\omega_1, \ldots, \omega_K$, the optimal solution to the above problem represents an optimal trade-off point between the cells' rates. Certainly, this point is Pareto-optimal, i.e., one cell cannot further improve its data rate without decreasing the data rate at (one or more) other cells. Similar to the optimization for the BC in a single-cell network (3.16), problem (3.26) is nonconvex. Thus, obtaining its globally optimal solution is a highly complex process. Fortunately, the iterative algorithms developed for the BC in [5, 38], as mentioned in Sect. 3.1.3, can be readily applied to this multicell problem. Alternately, by successively approximating the nonconvex part in the objective function of (3.26) to a convex lower bound function, the work in [19] proposed an algorithm that converges monotonically to a local optimal solution. However, these solution approaches are centralized and might not be suitable to the multicell setup. Recent works in [15, 35, 36] proposed an interference pricing mechanism that decomposes the nonconvex problem (3.26) and solves it on a per-cell basis. The idea of this approach is as follows. To isolate the ICI terms that render problem (3.26) nonconvex, we define the weighted sum-rate at all BSs except BS-q as $f_q(\mathbf{Q}_q, \mathbf{Q}_{-q}) = \sum_{r \neq q}^{Q} \omega_r R_r(\mathbf{Q}_q, \mathbf{Q}_{-q})$. At an instance of $(\mathbf{Q}_q, \mathbf{Q}_{-q})$, evaluated at

$(\bar{\mathbf{Q}}_q, \bar{\mathbf{Q}}_{-q})$, after taking the Taylor's expansion of $f_q(\cdot)$ and retaining only the linear term, the optimization (3.26) can be approximated by a set of Q per-cell problems

$$\underset{\mathbf{Q}_q}{\text{maximize}} \quad \omega_q \left| \mathbf{I} + \mathbf{H}_{qq}^H \mathbf{R}_q^{-1}(\mathbf{Q}_{-q})\mathbf{H}_{qq}\mathbf{Q}_q \right| - \text{Tr}\{\mathbf{A}_q\mathbf{Q}_q\} \qquad (3.28)$$

$$\text{subject to} \quad \text{Tr}\{\mathbf{Q}_q\} \leq P_q^{\max},$$

where $\mathbf{A}_q$, obtained from the negative derivative of $f_q(\cdot)$ with respect to $\mathbf{Q}_q$ [15], is the pricing matrix charged on the ICI caused by BS-q. Note that this approximated problem only requires local channel information pertained to BS-q. In addition, it is convex on $\mathbf{Q}_q$, and its optimal solution can easily be obtained in a closed-form WF solution [15]. Thus, the problem can be solved locally at each BS. Interestingly, the algorithm was shown to converge monotonically to a local optimal solution of the original problem (3.26). It is worth mentioning that while this solution approach can be implemented in a distributed manner, it still demands message passing to calculate and exchange the parameters $\mathbf{A}_q$'s.

When the design setting is to jointly minimize the power consumption, the joint optimization for CoMP under the IC mode can be stated as

$$\underset{\mathbf{Q}_1,\dots,\mathbf{Q}_K}{\text{minimize}} \quad \sum_{q=1}^{Q} \omega_q \text{Tr}\{\mathbf{Q}_q\} \qquad (3.29)$$

$$\text{subject to} \quad \left| \mathbf{I} + \mathbf{H}_{qq}^H \mathbf{R}_q^{-1}(\mathbf{Q}_{-q})\mathbf{H}_{qq}\mathbf{Q}_q \right| \geq R_q^{\min}, \ \forall q.$$

Similar to problem (3.26), this optimization is nonconvex, except for the MISO case, i.e, $N_q = 1, \forall q$. For that case, the work in [8] showed that the optimal beamformers (precoders) can be found in a distributed manner. Nonetheless, the distributed implementation of the algorithm proposed in [8] comes with several requirements, including perfect channel reciprocals, instant signaling exchanges, and synchronization among the coordinated BSs. To alleviate these drawbacks, we will consider a new game with pricing consideration that retains the advantages of the multicell game in CoMP with IA mode later in Chap. 4.

3.3   Concluding Remarks

In summary, this section has discussed various linear and nonlinear precoding designs in a single-cell network and examined how these designs have been adopted to the context of CoMP. The key aspect of precoding designs in a CoMP system is the consideration of the effect of ICI to the overall system performance. When CoMP is deployed under the IA mode, it is important to examine how each BS strategically adapts its precoding strategy accordingly to the amount of ICI at its connected MS. When CoMP is operating in the IC mode, the main concern is the joint control of the ICI by means of precoding in order to optimize the overall system performance.

It is worth mentioning that most of the related works in literature study the CoMP system with *one* MS per cell. Only a limited number of works, such as [8, 37, 42], considered a CoMP system with multiple MSs at each cell for the IC mode. It is observed that few works in literature have provided a thorough overview on how the ICI affects the multiuser precoding process in a CoMP system. In addition, while research on single-cell precoding techniques is quite plentiful, as previously mentioned in this section, many of these techniques do not have a counterpart version for a CoMP system in the presence of ICI. Under a more general setting with multiple MSs per cell, each BS has to take into account the resource allocation between its connected MSs and the signaling with other coordinated BSs in the system as well. In addition, each BS should be able to determine its precoders in fully distributed manner. How these factors impact on the precoding process at each coordinated BS are the main concerns of this monograph. Recent advances in precoding designs for a multiuser CoMP system in [20–23] will be covered intensively in this monograph.

References

1. Bengtsson, M., Ottersten, B.: Optimal and suboptimal transmit beamforming. L. C. Godara, ed., CRC Press (2001)
2. Bjornson, E., Zakhour, R., Gesbert, D., Ottersten, B.: Cooperative multicell precoding: Rate region characterization and distributed strategies with instantaneous and statistical CSI. IEEE Trans. Signal Process. **58**(9), 4298–4310 (2010)
3. Caire, G., Shamai, S.: On the achievable throughput of a multiantenna Gaussian broadcast channel. IEEE Trans. Inform. Theory **49**(7), 1691–1706 (2003)
4. Choi, L.U., Murch, R.: A transmit preprocessing technique for multiuser MIMO systems using a decomposition approach. IEEE Trans. Wireless Commun. **3**(1), 20–24 (2004)
5. Christensen, S.S., Argawal, R., E. de Carvalho, Cioffi, J.M.: Weighted sum-rate maximization using weighted MMSE for MIMO-BC beamforming design. IEEE Trans. Wireless Commun. **7**(12), 4792–4799 (2008)
6. Costa, M.: Writing on dirty paper. IEEE Trans. Inform. Theory **29**(3), 439–441 (1983)
7. Cover, T.M., Thomas, J.A.: Elements of Information Theory. John Wiley and Sons, Inc., New York: USA (1991)
8. Dahrouj, H., Yu, W.: Coordinated beamforming for the multicell multi-antenna wireless system. IEEE Trans. Wireless Commun. **9**(5), 1748–1759 (2010)
9. Dubey, P.: Inefficiency of Nash equilibria. Math. Oper. Res. **11**(1), 1–8 (1986)
10. Etkin, R., Parekh, A., Tse, D.N.C.: Spectrum sharing for unlicensed bands. IEEE J. Select. Areas in Commun. **25**(3), 517–528 (2007)
11. Foschini, G.J., Miljanic, Z.: A simple distributed autonomous power control algorithm and its convergence. IEEE Trans. Veh. Technol. **42**(4), 641–646 (1993)
12. Gohary, R., Yanikomeroglu, H.: Convergence of iterative water-filling with quantized feedback: A sufficient condition. IEEE Trans. Signal Process. **60**(5), 2688–2693 (2012)
13. Goodman, D.J., Mandayam, N.B.: Power control for wireless data. IEEE Pers. Commun. **7**, 48–54 (2000)

14. Jindal, N., Rhee, W., Vishwanath, S., Jafar, S.A., Goldsmith, A.: Sum power iterative water-filling for multi-antenna Gaussian broadcast channels. IEEE Trans. Inform. Theory **51**(4), 1570–1580 (2005)

15. Kim, S.J., Giannakis, G.B.: Optimal resource allocation for MIMO ad hoc cognitive radio networks. IEEE Trans. Inform. Theory **57**(5), 3117–3131 (2011)

16. Larsson, E., Jorswieck, E.: Competition versus cooperation on the MISO interference channel. IEEE J. Select. Areas in Commun. **26**(7), 1059–1069 (2008)

17. Luo, Z.Q., Pang, J.S.: Analysis of iterative waterfilling algorithm for multiuser power control in digital subscriber lines. EURASIP Journal on Advances in Signal Processing **2006**(1), 024,012 (2006). DOI 10.1155/ASP/2006/24012. URL http://asp.eurasipjournals.com/content/2006/1/024012

18. Luo, Z.Q., Yu, W.: An introduction to convex optimization for communications and signal processing. IEEE J. Select. Areas in Commun. **24**(8), 1426–1438 (2006)

19. Ng, C., Huang, H.: Linear precoding in cooperative MIMO cellular networks with limited coordination clusters. IEEE J. Select. Areas in Commun. **28**(9), 1446–1454 (2010)

20. Nguyen, D.H.N., Le-Ngoc, T.: Multiuser downlink beamforming in multicell wireless systems: A game theoretical approach. IEEE Trans. Signal Process. **59**(7), 3326–3338 (2011)

21. Nguyen, D.H.N., Le-Ngoc, T.: Sum-rate maximization in the multicell MIMO broadcast channel with interference coordination. IEEE Trans. Signal Process. **62**(6), 1501–1513 (2014)

22. Nguyen, D.H.N., Le-Ngoc, T.: Sum-rate maximization in the multicell MIMO multiple-access channel with interference coordination. IEEE Trans. Wireless Commun. **13**(1), 36–48 (2014)

23. Nguyen, D.H.N., Nguyen-Le, H., Le-Ngoc, T.: Block-diagonalization precoding in a multiuser multicell MIMO system: Competition and coordination. IEEE Trans. Wireless Commun. **13**(2), 968–981 (2014)

24. Pan, Z., Wong, K.K., Ng, T.S.: Generalized multiuser orthogonal space-division multiplexing. IEEE Trans. Wireless Commun. **3**(6), 1969–1973 (2004)

25. Pang, J.S., Scutari, G., Facchinei, F., Wang, C.: Distributed power allocation with rate constraints in Gaussian parallel interference channels. IEEE Trans. Inform. Theory **54**(8), 3471–3489 (2008)

26. Rashid-Farrokhi, F., Liu, K.J.R., Tassiulas, L.: Transmit beamforming and power control for cellular wireless systems. IEEE J. Select. Areas in Commun. **16**(8), 1437–1450 (1998)

27. Rashid-Farrokhi, F., Tassiulas, L., Liu, K.J.: Joint optimal power control and beamforming in wireless networks using antenna arrays. IEEE Trans. Commun. **46**(10), 1313–1323 (1998)

28. Saraydar, C.U., Mandayam, N.B., Goodman, D.J.: Efficient power control via pricing in wireless data networks. IEEE Trans. Commun. **50**(2), 291–303 (2002)

29. Schubert, M., Boche, H.: Solution of the multiuser downlink beamforming problem with individual SINR constraints. IEEE Trans. Veh. Technol. **53**(1), 18–28 (2004)

30. Scutari, G., Palomar, D.P., Barbarossa, S.: Asynchronous iterative waterffling for Gaussian frequency-selective interference channels. IEEE Trans. Inform. Theory **54**(7), 2868–2878 (2008)

31. Scutari, G., Palomar, D.P., Barbarossa, S.: Competitive design of multiuser MIMO system based on game theory: a unified view. IEEE J. Select. Areas in Commun. **26**(9), 1089–1102 (2008)

32. Scutari, G., Palomar, D.P., Barbarossa, S.: Optimal linear precoding strategies for wideband noncooperative systems based on game theory – Part I: Nash Equilibria. IEEE Trans. Signal Process. **56**(3), 1230–1249 (2008)

33. Scutari, G., Palomar, D.P., Barbarossa, S.: Optimal linear precoding strategies for wideband noncooperative systems based on game theory – Part II: Algorithms. IEEE Trans. Signal Process. **56**(3), 1250–1277 (2008)

34. Scutari, G., Palomar, D.P., Barbarossa, S.: The MIMO iterative waterfilling algorithm. IEEE Trans. Signal Process. **57**(5), 1917–1935 (2009)

35. Shi, C., Berry, R.A., Honig, M.L.: Monotonic convergence of distributed interference pricing in wireless networks. In: Proc. IEEE Int. Symp. Inform. Theory, pp. 1619–1623. Seoul, Republic of Korea (2009)

36. Shi, C., Schmidt, D.A., Berry, R.A., Honig, M.L., Utschick, W.: Distributed interference pricing for the MIMO interference channel. In: Proc. IEEE Int. Conf. Commun., pp. 1–5. Dresden, Germany (2009)

37. Shi, Q., Razaviyayn, M., Luo, Z.Q., He, C.: An iteratively weighted MMSE approach to distributed sum-utility maximization for MIMO interfering broadcast channel. IEEE Trans. Signal Process. **59**(9), 4331–4340 (2011)

38. Shi, S., Schubert, M., Boche, H.: Rate optimization for multiuser MIMO with linear processing. IEEE Trans. Signal Process. **56**(8), 4020–4030 (2008)

39. Shum, K., Leung, K.K., Sung, C.W.: Convergence of iterative waterfilling algorithm for Gaussian interference channels. IEEE J. Select. Areas in Commun. **25**(6), 1091–1100 (2007)

40. Spencer, Q., Swindlehurst, A., Haardt, M.: Zero-forcing methods for downlink spatial multiplexing in multiuser MIMO channels. IEEE Trans. Signal Process. **52**(2), 461–471 (2004)

41. Telatar, I.E.: Capacity of multi-antenna Gaussian channels. Eur. Trans. Telecommu. **10**, 585–595 (1999)

42. Venturino, L., Prasad, N., Wang, X.: Coordinated linear beamforming in downlink multi-cell wireless networks. IEEE Trans. Wireless Commun. **9**(4), 1451ï£¡–1461 (2010)

43. Vishwanath, S., Jindal, N., Goldsmith, A.: Duality, achievable rates and sum-rate capacity of Gaussian MIMO broadcast channels. IEEE Trans. Inform. Theory **49**(10), 2658–2668 (2003)

44. Visotsky, E., Madhow, U.: Optimum beamforming using transmit antenna arrays. In: Proc. IEEE Veh. Technol. Conf., vol. 1, pp. 851–856 (1999)

45. Viswanath, P., Tse, D.: Sum capacity of the vector Gaussian broadcast channel and uplink-downlink duality. IEEE Trans. Inform. Theory **49**(8), 1912–1921 (2003)

46. Wiesel, A., Eldar, Y.C., Shamai, S.: Linear precoding via conic optimization for fixed MIMO receivers. IEEE Trans. Signal Process. **54**(1), 161–176 (2006)

47. Wong, K.K., Murch, R., Letaief, K.: A joint-channel diagonalization for multiuser MIMO antenna systems. IEEE Trans. Wireless Commun. **2**(4), 773–786 (2003)

48. Yates, R.D.: A framework for uplink power control in cellular radio systems. IEEE J. Select. Areas in Commun. **13**(7), 1341–1347 (1995)

49. Yu, W.: Sum-capacity computation for the Gaussian vector broadcast channel via dual decomposition. IEEE Trans. Inform. Theory **52**(2), 754–759 (2006)

50. Yu, W., Cioffi, J.: Sum capacity of Gaussian vector broadcast channels. IEEE Trans. Inform. Theory **50**(9), 1875–1892 (2004)

51. Yu, W., Ginis, G., Cioffi, J.M.: Distributed multiuser power control for digital subscriber lines. IEEE J. Select. Areas in Commun. **20**(5), 1105–1115 (2002)

52. Yu, W., Lan, T.: Transmitter optimization for the multi-antenna downlink with per-antenna power constraints. IEEE Trans. Signal Process. **55**(6), 2646–2660 (2007)

53. Yu, W., Rhee, W., Boyd, S., Cioffi, J.M.: Iterative water-filing for Gaussian multiple-access channels. IEEE Trans. Inform. Theory **50**(1), 145–152 (2004)

Chapter 4
Multiuser Downlink Beamforming in Multicell Wireless Systems

This chapter considers a game theoretical approach to study the competitive precoding design in a multiuser multicell system, where each BS concurrently serves multiple MSs (or users). Sharing the same frequency band, the BS of each cell wishes to design the optimal downlink beamformers for its users in order to minimize its transmit power, given a set of target signal-to-interference-plus-noise ratios (SINRs) for the users in its cell. Under a similar setup, the work in [12] studied scheduling schemes to handle the ICI and provided a quality of service (QoS) guaranteed in the form of packet error rate. Multicell downlink beamforming with coordination was considered in [5], where the total weighted transmit power across multiple BSs is jointly minimized. Via the concept of uplink-downlink duality, it is shown in [5] that such a jointly optimal design can be implemented in a distributed manner under certain requirements, including perfect channel reciprocal from each BS to each MS (not necessarily in the same cell) and synchronization among the BSs. These requirements, which may be difficult to meet in practice, are the drawbacks of the distributed implementation in the coordinated design. Conversely, in the competitive design, where the multicell beamformers are devised on per-cell basis with no centralized control, these requirements can be alleviated.

Using the game-theory framework, we establish the best response strategy of a cell, given the beamforming strategies from other cells. Then, it is shown that such best response strategy is a *standard* function [19], which guarantees the uniqueness of the NE and the convergence of the distributed algorithm. This is the distinction of this power minimization game, compared to typical n-person concave games in an OFDM system with WF as the optimal strategy [9, 14, 15, 17, 20]. In addition, necessary and sufficient conditions for the existence of the NE are also given. A comparison to the fully coordinated multicell downlink beamforming design is then presented in this chapter.

It is worth mentioning that the NE of the multicell game needs not to be Pareto-optimal, i.e., it may not stay in the surface established by the coordinated design. Moreover, it may happen that a beamformer design in one cell is highly correlated to the channel of the other cells, which then causes significant ICI.

D.H.N. Nguyen and T. Le-Ngoc, *Wireless Coordinated Multicell Systems: Architectures and Precoding Designs*, SpringerBriefs in Computer Science, DOI 10.1007/978-3-319-06337-9_4, © The Author(s) 2014

To avoid this undesired effect, we consider a new multicell downlink beamforming game with pricing consideration, where each BS voluntarily attempts to minimize the interference induced to other cells. This pricing technique allows a BS to steer its beamformers in a more cooperative way, which results in a more Pareto-efficient NE. The characterization of the new game's NE reveals that under certain conditions, the new NE point is able to approach the performance established by the coordinated design, while retaining the distributed nature of the SNG.

4.1 System Model

Fig. 4.1 An example of a multicell system with 3 base-stations and 2 users per cell

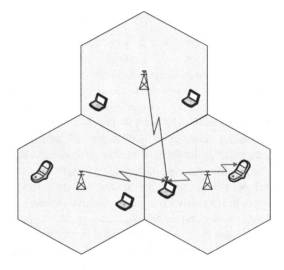

Consider a multiuser downlink beamforming system with Q separate cells operating on the same frequency channel, as illustrated in Fig. 4.1. In each cell, one multiple-antenna BS is concurrently sending independent information to several remote single-antenna MSs. Let $\Omega = \{1, \ldots, Q\}$ denote the set of the cells (players), and define $\Omega_{-q} = \Omega \backslash \{q\}$. For simplicity of presentation, it is assumed that each BS is equipped with M antennas, and is serving K MSs. Note that each user (MS) is now subject to the co-channel interference from other cells (ICI), in addition to the interference caused by the signals intended for other users in the same cell (intra-cell interference). In a competitive design for this multicell system, it is assumed that each BS has full knowledge of the downlink channels within its cell, but not the inter-cell channels. Thus, each BS is only able to manage the intra-cell interference. On the other hand, the BS treats the ICI as background noise. In the later parts of this chapter, it is assumed that a BS may possess the full or partial channel information to the users in other cells. The additional channel knowledge then allows a BS to control the ICI as well.

Considering the transmission at a particular cell, say cell-q, its downlink channel can be modeled as

$$y_{q_i} = \mathbf{h}_{qq_i}^H \mathbf{x}_q + \sum_{m \neq q}^{Q} \mathbf{h}_{mq_i}^H \mathbf{x}_m + z_{q_i}, \qquad (4.1)$$

where $\mathbf{x}_m$ is an $M \times 1$ complex vector representing the transmitted signal at BS-m, $\mathbf{h}_{mq_i}$ is an $M \times 1$ complex channel vector from BS-m to user-i of cell-q, y_{q_i} represents the received signal at user-i, and z_{q_i} is $\mathscr{CN}(0, \sigma^2)$. It is assumed that the channel vector $\mathbf{h}_{qq_i}^*$ is known at both the BS and user-i of cell-q, whereas the cross-cell channel $\mathbf{h}_{mq_i}, m \neq q$ is unknown.

In a beamforming design, the transmitted signal $\mathbf{x}_q$ is of the form

$$\mathbf{x}_q = \sum_{i=1}^{K} x_{q_i} \mathbf{w}_{q_i}, \qquad (4.2)$$

where x_{q_i} is a complex scalar representing the signal intended for user-i, and $\mathbf{w}_{q_i}$ is an $M \times 1$ beamforming vector for user-i. Without loss of generality, let $\mathbb{E}[|x_{q_i}|] = 1$. It is easy to verify that the SINR at user-i of cell-q is

$$\mathrm{SINR}_{q_i} = \frac{\left|\mathbf{w}_{q_i}^H \mathbf{h}_{qq_i}\right|^2}{\sum_{j \neq i}^{K} \left|\mathbf{w}_{q_j}^H \mathbf{h}_{qq_i}\right|^2 + \sum_{m \neq q}^{Q} \sum_{j=1}^{K} \left|\mathbf{w}_{m_j}^H \mathbf{h}_{mq_i}\right|^2 + \sigma^2}. \qquad (4.3)$$

Note that the received signal at user-i of cell-q is corrupted by the intra-cell interference $\sum_{j \neq i}^{K} \left|\mathbf{w}_{q_j}^H \mathbf{h}_{qq_i}\right|^2$, the ICI $\sum_{m \neq q}^{Q} \sum_{j=1}^{K} \left|\mathbf{w}_{m_j}^H \mathbf{h}_{mq_i}\right|^2$, as well as the AWGN. Although the channel state information from other cells is not known at both the users and BS of cell-q, each user can measure its total interference and report back to the BS. The BS, having known the channel to the users in its cell, can determine the total ICI plus AWGN at each user.

4.2 Multicell Downlink Beamforming Game

4.2.1 Problem Formulation

In the first part of this chapter is concerned with formulating the multicell downlink beamforming design within the framework of game theory. In particular, consider the multicell system as a SNG, where the players are the cells and the payoff functions are the transmit powers of the BSs. More specifically, each player competes with each other by choosing the downlink beamformer design that greedily minimizes its own transmit power subject to a given set of target SINRs at the users within its cell. Each channel is assumed to vary sufficiently slowly such that it can be considered fixed while the game is being played.

Define the precoding matrix $\mathbf{W}_q = [\mathbf{w}_{q_1}, \ldots, \mathbf{w}_{q_K}]$ as the strategy at BS-q, and $\mathbf{W}_{-q}$ as the precoding strategy of all the BSs, except BS-q. The transmit power at BS-q, is then given by $\|\mathbf{W}_q\|_F^2$. Further define the set of admissible beamforming strategies $\mathbf{W}_q \in \mathscr{P}_q(\mathbf{W}_{-q})$ of cell-q as

$$\mathscr{P}_q(\mathbf{W}_{-q}) = \left\{\mathbf{W}_q \in \mathbb{C}^{M \times K} : \text{SINR}_{q_i}(\mathbf{W}_q, \mathbf{W}_{-q}) \geq \gamma_{q_i}, \forall i \right\},$$

where γ_{q_i} is the target SINR at user-i of cell-q.

At cell-q, denote the total ICI plus background noise (IPN) at the ith user as $r_{-q_i}(\mathbf{W}_{-q}) = \sum_{m \neq q}^{Q} \sum_{j=1}^{K} \left|\mathbf{w}_{m_j}^H \mathbf{h}_{m q_i}\right|^2 + \sigma^2 = \sum_{m \neq q}^{Q} \left\|\mathbf{W}_m^H \mathbf{h}_{m q_i}\right\|^2 + \sigma^2$. Furthermore, denote $\mathbf{r}_{-q} = [r_{-q_1}, \ldots, r_{-q_K}]^T$. Note that the set of feasible strategies $\mathscr{P}(\mathbf{W}_{-q})$ of cell-q depends on the beamforming strategies $\mathbf{W}_{-q}$ of all the other cells. Mathematically, the corresponding game has the following structure

$$\mathscr{G} = \left(\Omega, \ \left\{\mathscr{P}_q(\mathbf{W}_{-q})\right\}_{q \in \Omega}, \ \left\{t_q(\mathbf{W}_q)\right\}_{q \in \Omega}\right),$$

where $t_q(\mathbf{W}_q) = \|\mathbf{W}_q\|_F^2$ is the transmit power at BS-q.[1] Given the beamforming design of the others, reflected by the IPN vector $\mathbf{r}_{-q}$, the optimal or best response strategy of the qth BS is the solution to the following optimization problem

$$\underset{\mathbf{W}_q}{\text{minimize}} \ \ \|\mathbf{W}_q\|_F^2 \tag{4.4}$$

$$\text{subject to} \ \ \frac{\left|\mathbf{w}_{q_i}^H \mathbf{h}_{q q_i}\right|^2}{\sum_{j \neq i}^{K} \left|\mathbf{w}_{q_j}^H \mathbf{h}_{q q_i}\right|^2 + r_{-q_i}} \geq \gamma_{q_i}, \ \forall i.$$

It is noted that there are several numerical approaches to find the optimal solution to this downlink beamforming problem, as mentioned in Chap. 3. In a multicell configuration, the problem arisen here is when one player changes its beamforming matrix, the other players also need to change their own beamforming matrices in order to achieve its target SINRs. Our interest is to investigate whether game $\mathscr{G}$ eventually converges into a stable point, i.e., an NE; and if an NE exists, whether its uniqueness hold. A feasible strategy profile $\mathbf{W}^\star = \{\mathbf{W}_q^\star\}_{q=1}^{Q}$ is an NE of game $\mathscr{G}$ if

$$t_q(\mathbf{W}_q^\star) \leq t_q(\mathbf{W}_q), \ \forall \mathbf{W}_q \in \mathscr{P}_q(\mathbf{W}_{-q}^\star), \quad \forall q \in \Omega. \tag{4.5}$$

---

[1] According to recent use, this game may be referred to as a generalized Nash equilibrium problem where the admissible strategy set of a player depends on the other players' strategy [11].

At the NE point, given the beamforming matrices from other cells, a BS does not have the incentive to unilaterally change its own beamforming matrix, i.e., it will consume more power to obtain the same SINR targets. In the following sections, by first studying the best response strategy of each player, the NE of game $\mathscr{G}$ is subsequently characterized.

4.2.2 Best Response Strategy

This section begins with some related propositions to simplify the analysis of the game and characterize the best response strategy.

Proposition 4.1. *If $\mathbf{W}_q^\star$ is the optimal beamforming strategy for cell-q, then $\mathbf{W}_q^\star\mathbf{R}$, where $\mathbf{R} = \text{diag}\left(e^{j\theta_1}, \ldots, e^{j\theta_K}\right)$, $\forall\theta_1, \ldots, \theta_K$ is also optimal.*

This proposition stems from the fact that the effective SINR at each user is invariant to a constant phase change of the beamforming vector of any other user.

Proposition 4.2. *With unlimited transmit power, the feasibility of the optimization problem (4.4) at cell-q is only dependent on the channel $\mathbf{h}_{qq_1}, \ldots, \mathbf{h}_{qq_K}$ and the set of target SINRs $\gamma_{q_1}, \ldots, \gamma_{q_K}$. It is independent of the IPN vector $\mathbf{r}_{-q} > 0$.*

This proposition comes directly from Proposition 1 in [18], which shows that the rank of the matrix $\mathbf{H}_q = [\mathbf{h}_{qq_1}, \ldots, \mathbf{h}_{qq_K}]$ determines the feasibility of the optimization problem (4.4). In addition, when $\mathbf{H}_q$ is full-rank, any set of target SINRs $\gamma_{q_1}, \ldots, \gamma_{q_K}$ is feasible. In this system model, it is assumed that $\mathbf{H}_q$ is full-rank, $\forall q$.

Proposition 4.3. *For two different IPN vectors at cell-q, $\mathbf{r}_{-q}$ and $\bar{\mathbf{r}}_{-q}$, the optimal beamforming vector $\mathbf{w}_{q_i}^\star$, corresponding to $\mathbf{r}_{-q}$, is a scaled version of the optimal beamforming vector $\bar{\mathbf{w}}_{q_i}^\star$, corresponding to $\bar{\mathbf{r}}_{-q}$.*

Proof. It is first observed that the IPN components in $\mathbf{r}_{-q}$ are scalars, which therefore should not impact on the directions of the beamformers at BS-q. To support this observation, we revisit the dual uplink problem of problem (4.4), which turns out to be [21]

$$\underset{\lambda_{q_1}, \ldots, \lambda_{q_K}}{\text{maximize}} \sum_{i=1}^{K} \lambda_{q_i} r_{-q_i} \tag{4.6}$$

$$\text{subject to} \sum_{j \neq i}^{K} \lambda_{q_j} \mathbf{h}_{qq_j} \mathbf{h}_{qq_j}^H + \mathbf{I} \succeq \frac{\lambda_{q_i}}{\gamma_{q_i}} \mathbf{h}_{qq_i} \mathbf{h}_{qq_i}^H, \ \forall i.$$

This optimization is then equivalent to the dual uplink problem [21]

$$\underset{\substack{\lambda_{q_1},...,\lambda_{q_K} \\ \mathbf{w}_{q_1},...,\mathbf{w}_{q_K}}}{\text{minimize}} \quad \sum_{i=1}^{K} \lambda_{q_i} r_{-q_i} \tag{4.7}$$

$$\text{subject to} \quad \frac{\lambda_{q_i} \left|\hat{\mathbf{w}}_{q_i}^H \mathbf{h}_{qq_i}\right|^2}{\sum_{j \neq i}^{K} \lambda_{q_j} \left|\hat{\mathbf{w}}_{q_i}^H \mathbf{h}_{qq_j}\right|^2 + \hat{\mathbf{w}}_{q_i}^H \hat{\mathbf{w}}_{q_i}} \geq \gamma_{q_i}, \ \forall i,$$

where the optimization is over a weighted sum-power of the uplink powers λ_{q_i}'s and the receive beamformer vectors $\hat{\mathbf{w}}_{q_i}$'s. It is to be noted that the optimal uplink power $\lambda_{q_i}^\star$ is only dictated by the constraints, but not the objective function. As the constraints are met with equality at optimality, the solution of λ_{q_i} can be obtained by the fixed point iteration [19]

$$\lambda_{q_i}^{(n+1)} = \frac{\gamma_{q_i}}{1 + \gamma_{q_i}} \times \frac{1}{\mathbf{h}_{q_i}^H \left(\sum_{j=1}^{K} \lambda_{q_j}^{(n)} \mathbf{h}_{qq_j} \mathbf{h}_{qq_j}^H + \mathbf{I}\right)^{-1} \mathbf{h}_{q_i}}. \tag{4.8}$$

The optimal receive beamformer vector $\hat{\mathbf{w}}_{q_i}^\star$ is the minimum mean square error (MMSE) receiver, i.e., $\hat{\mathbf{w}}_{q_i}^\star = \left(\sum_{j=1}^{K} \lambda_{q_j}^\star \mathbf{h}_{qq_j} \mathbf{h}_{qq_j}^H + \mathbf{I}\right)^{-1} \mathbf{h}_{qq_i}$. Clearly, the optimal uplink power $\lambda_{q_i}^\star$ and the optimal receive beamformer $\hat{\mathbf{w}}_{q_i}^\star$ are independent of r_{-q_i}. The optimal downlink beamformer vectors $\mathbf{w}_{q_i}^\star$ then can be found to be a scaled version of the uplink receive beamformer $\hat{\mathbf{w}}_{q_i}^\star$ [18]. Thus, both the optimal beamforming vector $\mathbf{w}_{q_i}^\star$, corresponding to $\mathbf{r}_{-q}$, and the vector $\bar{\mathbf{w}}_{q_i}^\star$, corresponding to $\bar{\mathbf{r}}_{-q}$, are scaled versions of $\hat{\mathbf{w}}_{q_i}$. This concludes the proof for this proposition.

From Proposition 4.3, it is to be noted that whenever the IPN vector $\mathbf{r}_{-q}$ is changed, BS-q only needs to adjust the allocated power for each user, but not the beam patterns to its users.[2] Thus, BS-q can determine its beam pattern first, and then allocate the appropriate power to each user subject to the ICI in $\mathbf{r}_{-q}$.

The optimal beam patterns at each cell can be determined by solving problem (4.4) in the absence of ICI, which is

$$\underset{\mathbf{W}_q}{\text{minimize}} \quad \sum_{i=1}^{K} \left\|\mathbf{w}_{q_i}\right\|^2 \tag{4.9}$$

$$\text{subject to} \quad \frac{\left|\mathbf{w}_{q_i}^H \mathbf{h}_{qq_i}\right|^2}{\sum_{j \neq i}^{K} \left|\mathbf{w}_{q_j}^H \mathbf{h}_{qq_i}\right|^2 + \sigma^2} \geq \gamma_{q_i}, \ \forall i.$$

This problem can be easily solved by the techniques mentioned in the previous section. The beamforming pattern for each user in cell-q is determined as $\tilde{\mathbf{w}}_{q_i} = \mathbf{w}_{q_i} / \left\|\mathbf{w}_{q_i}\right\|$. By Proposition 4.3, problem (4.4) can be restated as

---

[2]The beam pattern is defined as the norm-1 $\mathbf{w}_{q_i} / \left\|\mathbf{w}_{q_i}\right\|$.

$$\underset{p_{q_1},\dots,p_{q_K}}{\text{minimize}} \quad \sum_{i=1}^{K} p_{q_i} \tag{4.10}$$

$$\text{subject to} \quad \frac{p_{q_i} \left| \tilde{\mathbf{w}}_{q_i}^H \mathbf{h}_{qq_i} \right|^2}{\sum_{j \neq i}^{K} p_{q_j} \left| \tilde{\mathbf{w}}_{q_j}^H \mathbf{h}_{qq_i} \right|^2 + r_{-q_i}} \geq \gamma_{q_i}, \ \forall i,$$

$$p_i \geq 0, \ \forall i, \tag{4.11}$$

which is reduced back to a power allocation problem, where p_{q_i} is the allocated power for user-i. Denote $\mathbf{p}_q = (p_{q_i}, \dots, p_{q_K})$, $\mathbf{p} = (\mathbf{p}_1, \dots, \mathbf{p}_Q)$, and $\mathbf{p}_{-q} = (\mathbf{p}_1, \dots, \mathbf{p}_{q-1}, \mathbf{p}_{q+1}, \dots, \mathbf{p}_Q)$. Note that the strategy set of player-q is redefined as

$$\mathscr{P}_q(\mathbf{p}_{-q}) = \left\{ \mathbf{p}_q \in \mathbb{R}_+^K : \text{SINR}_{q_i}(\mathbf{p}_q, \mathbf{p}_{-q}) \geq \gamma_{q_i}, \ \forall i \right\}. \tag{4.12}$$

After obtaining the optimal solution $p_{q_i}^{\star}$ of problem (4.10), the optimal beamforming vectors corresponding to $\mathbf{r}_{-q}$ are $\sqrt{p_{q_i}^{\star}} \tilde{\mathbf{w}}_{q_i}$. The following proposition presents the analytical solution to (4.10).

Proposition 4.4. *The optimal solution of (4.10),* $\mathbf{p}_q^{\star} = [p_{q_1}^{\star}, \dots, p_{q_K}^{\star}]^T$, *is given by*

$$\mathbf{p}_q^{\star} = \mathbf{G}_{qq}^{-1} \mathbf{r}_{-q}, \tag{4.13}$$

where $\mathbf{G}_{qq} \in \mathbb{R}^{K \times K}$, *defined as* $[\mathbf{G}_{qq}]_{i,i} = (1/\gamma_{q_i}) \left| \tilde{\mathbf{w}}_{q_i}^H \mathbf{h}_{qq_i} \right|^2$ *and* $[\mathbf{G}_{qq}]_{i,j} = -\left| \tilde{\mathbf{w}}_{q_j}^H \mathbf{h}_{qq_i} \right|^2$ *if* $i \neq j$, *is invertible. Furthermore,* $\mathbf{p}_q^{\star} > 0, \ \forall \mathbf{r}_{-q} > 0$.

Proof. Since all the SINR constraints in (4.10) are met with equality at optimality, they are equivalent to

$$p_{q_i}^{\star} \frac{\left| \tilde{\mathbf{w}}_{q_i}^H \mathbf{h}_{qq_i} \right|^2}{\gamma_{q_i}} - \sum_{j \neq i}^{K} p_{q_j}^{\star} \left| \tilde{\mathbf{w}}_{q_j}^H \mathbf{h}_{qq_i} \right|^2 = r_{-q_i}, \ i = 1, \dots, K.$$

The solution of this set of K equations with K variables are the optimal solution of (4.10). Rewriting these K equations in matrix form, one has $\mathbf{G}_{qq} \mathbf{p}_q^{\star} = \mathbf{r}_{-q}$. From Proposition 4.2, as the problem (4.10) is always feasible $\forall \mathbf{r}_{-q} > 0$, $\mathbf{G}_{qq}$ has to be invertible, and $\mathbf{p}_q^{\star} = \mathbf{G}_{qq}^{-1} \mathbf{r}_{-q} > 0$ is uniquely defined, $\forall \mathbf{r}_{-q} > 0$.

Denote $\mathbf{G}_{mq} \in \mathbb{R}^{K \times K}, m \neq q$ as the ICI matrix from cell-m to cell-q, where $[\mathbf{G}_{mq}]_{i,j} = \left| \tilde{\mathbf{w}}_{m_j}^H \mathbf{h}_{mq_i} \right|^2$. Then $\mathbf{G}_{mq} \mathbf{p}_m$ is the interference vector caused by BS-m to the K users of cell-q. Thus, one has $\mathbf{r}_{-q} = \sum_{m \neq q}^{Q} \mathbf{G}_{mq} \mathbf{p}_m + \mathbf{1}\sigma^2$.

From Proposition 4.4, the best response strategy of the qth cell subject to the strategy of Ω_{-q} is

$$\mathbf{p}_q^\star = \mathrm{BR}_q\left(\mathbf{p}_{-q}\right) = \mathbf{G}_{qq}^{-1}\left(\sum_{m\neq q}^{Q} \mathbf{G}_{mq}\mathbf{p}_m + \mathbf{1}\sigma^2\right), \ \forall q. \tag{4.14}$$

The NEs of game $\mathscr{G}$ can now be redefined as the intersection points of the BRs, i.e.,

$$\mathbf{p}_q^\star = \mathbf{G}_{qq}^{-1}\left(\sum_{m\neq q}^{Q} \mathbf{G}_{mq}\mathbf{p}_m^\star + \mathbf{1}\sigma^2\right), \ \forall q. \tag{4.15}$$

The next lemma shows that the best response function in (4.14) is *standard* [19], which guarantees the uniqueness of the NE if such NE exists [8].

Lemma 4.1. *The best response function is a standard function.*

Proof. First, define $\mathbf{p} = [\mathbf{p}_1^T, \ldots, \mathbf{p}_Q^T]^T$, then $\mathrm{BR}_q(\mathbf{p}) \triangleq \mathrm{BR}_q(\mathbf{p}_{-q})$. It is needed to show that the best response function $\mathrm{BR}_q(\mathbf{p})$ meets the three requirements of a *standard* function:

1. *Positivity:* for any $\mathbf{p} \geq 0$, as $\mathbf{G}_{mq}$ is a positive matrix, one has $\sum_{m\neq q}^{Q} \mathbf{G}_{mq}\mathbf{p}_m + \mathbf{1}\sigma^2 > 0, \forall q$. Thus, $\mathrm{BR}_q(\mathbf{p}) = \mathbf{G}_{qq}^{-1}\left(\sum_{m\neq q}^{Q} \mathbf{G}_{mq}\mathbf{p}_m + \mathbf{1}\sigma^2\right) > 0, \forall q$, as a result from Proposition 4.4.

2. *Monotonicity:* for $\mathbf{p} \geq \mathbf{p}'$, then

$$\mathrm{BR}_q(\mathbf{p}) - \mathrm{BR}_q(\mathbf{p}') = \mathbf{G}_{qq}^{-1}\left[\sum_{m\neq q}^{Q} \mathbf{G}_{mq}\left(\mathbf{p}_m - \mathbf{p}_m'\right)\right] \geq 0, \ \forall q$$

   as a result from Proposition 4.4 and each $\mathbf{p}_m - \mathbf{p}_m' \geq 0$.

3. *Scalability:* $\forall \varepsilon > 1, \forall \mathbf{p} \geq 0$, one has

$$\varepsilon \mathrm{BR}_q(\mathbf{p}) - \mathrm{BR}_q(\varepsilon\mathbf{p}) = \varepsilon\mathbf{G}_{qq}^{-1}\mathbf{1}\sigma^2 - \mathbf{G}_{qq}^{-1}\mathbf{1}\sigma^2 = (\varepsilon - 1)\mathbf{G}_{qq}^{-1}\mathbf{1}\sigma^2 > 0.$$

Since $\mathrm{BR}_q(\mathbf{p}), q = 1, \ldots, Q$, are *standard* functions, the iteration $\mathbf{p}_q^{(t+1)} = \mathrm{BR}_q(\mathbf{p}^{(t)}), q = 1, \ldots, Q$, will converge from any starting point $\mathbf{p}^{(0)} > 0$ to a unique fixed point (when it exists) [19], which is the NE of game $\mathscr{G}$ [8]. In addition, the iterative update can be implemented in a fully asynchronous manner among the cells. It is to be noted that due to the monotonicity of the standard function, if the fixed point does not exist, the transmit power at each BS will increase without bound. To this end, one sufficient condition and one necessary condition guaranteeing the boundedness of the NE are examined. Such conditions equivalently guarantee the existence and uniqueness of the game's NE.

4.2.3 NE Existence and Uniqueness

4.2.3.1 A Sufficient Condition

This section is to examine the best response dynamic of game $\mathscr{G}$ as a mapping process. A sufficient condition is then presented such that the mapping is a contraction, which guarantees the existence of the fixed-point of the mapping, i.e., the NE of game $\mathscr{G}$. The obtained result is summarized in the following proposition.

Proposition 4.5. *The NE of game $\mathscr{G}$ exists if the spectral radius*

$$(C): \quad \varrho(\mathbf{S}) < 1, \tag{4.16}$$

where the square matrix $\mathbf{S} \in \mathbb{R}^{Q \times Q}$ is defined as

$$[\mathbf{S}]_{q,m} = \begin{cases} 0, & \text{if } m = q \\ \left\| \mathbf{G}_{qq}^{-1} \mathbf{G}_{mq} \right\|_F, & \text{if } m \neq q. \end{cases} \tag{4.17}$$

Proof. Let $\mathbf{T}_q(\mathbf{p}) = BR_q(\mathbf{p})$ and $\mathbf{T}(\mathbf{p}) = \left(BR_q(\mathbf{p})\right)_{q \in \Omega}$. Since $\mathbf{T}_q(\mathbf{p})$ is a mapping from $\mathbb{R}_+^{KQ}$ onto $\mathscr{P}_q(\mathbf{p}_{-q})$, which is a subset of $\mathbb{R}_+^K$, $\mathbf{T}(\mathbf{p})$ is a mapping from $\mathbb{R}_+^{KQ}$ onto a Cartesian product of Q $\mathbb{R}_+^K$ sets, i.e., $\mathbb{R}_+^{KQ}$. For some $\mathbf{w} = [w_1, \ldots, w_Q]^T > 0$, the mapping $\mathbf{T}(\mathbf{p})$ is a block-contraction of modulus α, with respect to the norm $\| \cdot \|_{\infty,\text{block}}^{\mathbf{w}}$, if there exists a non-negative constant $\alpha < 1$ such that

$$\left\| \mathbf{T}(\mathbf{p}) - \mathbf{T}(\mathbf{p}') \right\|_{2,\text{block}}^{\mathbf{w}} \leq \alpha \left\| \mathbf{p} - \mathbf{p}' \right\|_{2,\text{block}}^{\mathbf{w}}, \quad \forall \mathbf{p}, \mathbf{p}' \geq 0. \tag{4.18}$$

The condition $\alpha < 1$ is sufficient to guarantee the existence and uniqueness of the fixed point $\mathbf{p} = \mathbf{T}(\mathbf{p})$ as well as the convergence of the mapping to the fixed point. It is worth mentioning that this property has been commonly applied to the analysis of games with best response dynamics, such as the well-known IWF game in a multi-channel system [9, 14–16]. To this end, this contraction mapping property is exploited to establish a sufficient condition on the boundedness of the power update in game $\mathscr{G}$.

Let $e_{\mathbf{T}_q} = \left\| \mathbf{T}_q(\mathbf{p}) - \mathbf{T}_q(\mathbf{p}') \right\|_2$, and $e_q = \left\| \mathbf{p}_q - \mathbf{p}_q' \right\|_2$, then

$$\begin{aligned} e_{\mathbf{T}_q} &= \left\| BR_q(\mathbf{p}) - BR_q(\mathbf{p}') \right\|_2 \\ &= \left\| \mathbf{G}_q^{-1} \left[\sum_{m \neq q}^{Q} \mathbf{G}_{mq} (\mathbf{p}_m - \mathbf{p}_m') \right] \right\|_2 \\ &\leq \sum_{m \neq q}^{Q} \left\| \mathbf{G}_q^{-1} \mathbf{G}_{mq} \right\| \left\| \mathbf{p}_m - \mathbf{p}_m' \right\|_2, \end{aligned} \tag{4.19}$$

where the inequality is satisfied if the matrix norm $\| \cdot \|$ applied to $\mathbf{G}_q^{-1}\mathbf{G}_{mq}$ is consistent [7]. Here, the Frobenius norm $\| \cdot \|_F$ can be utilized since it is consistent and easy to compute. Define the vectors $\mathbf{e_T} = [e_{T_1}, \ldots, e_{T_Q}]^T$ and $\mathbf{e} = [e_1, \ldots, e_Q]^T$. Furthermore, define the square matrix $\mathbf{S} \in \mathbb{R}^{Q \times Q}$, where

$$[\mathbf{S}]_{q,m} = \begin{cases} 0, & \text{if } m = q \\ \left\| \mathbf{G}_{qq}^{-1} \mathbf{G}_{mq} \right\|_F, & \text{if } m \neq q \end{cases} . \tag{4.20}$$

Then, one has

$$\mathbf{e_T} \leq \mathbf{Se}. \tag{4.21}$$

Thus,

$$\|\mathbf{e_T}\|_{\infty,\text{vec}}^{\mathbf{w}} \leq \|\mathbf{Se}\|_{\infty,\text{vec}}^{\mathbf{w}} \leq \|\mathbf{S}\|_{\infty,\text{mat}}^{\mathbf{w}} \|\mathbf{e}\|_{\infty,\text{vec}}^{\mathbf{w}}, \tag{4.22}$$

as the induced ∞-norm $\| \cdot \|_{\infty,\text{mat}}^{\mathbf{w}}$ is consistent [7]. Then, one has

$$\begin{aligned} \left\| \mathbf{T}(\mathbf{p}) - \mathbf{T}(\mathbf{p}') \right\|_{2,\text{block}}^{\mathbf{w}} &= \max_{q \in \Omega} \frac{\|\text{BR}_q(\mathbf{p}) - \text{BR}_q(\mathbf{p}')\|_2}{w_q} \\ &= \|\mathbf{e_T}\|_{\infty,\text{vec}}^{\mathbf{w}} \\ &\leq \|\mathbf{S}\|_{\infty,\text{mat}}^{\mathbf{w}} \|\mathbf{e}\|_{\infty,\text{vec}}^{\mathbf{w}} \\ &= \|\mathbf{S}\|_{\infty,\text{mat}}^{\mathbf{w}} \left\| \mathbf{p} - \mathbf{p}' \right\|_{2,\text{block}}^{\mathbf{w}} . \end{aligned} \tag{4.23}$$

Thus, if $\|\mathbf{S}\|_{\infty,\text{mat}}^{\mathbf{w}} < 1$, the existence and the convergence to the fixed point are guaranteed because the mapping in (4.23) is a contraction. It is to be noted that if $\mathbf{S}$ is a non-negative matrix, there exists a positive vector $\mathbf{w}$ such that [3]

$$\|\mathbf{S}\|_{\infty,\text{mat}}^{\mathbf{w}} < 1 \quad \Longleftrightarrow \quad \varrho(\mathbf{S}) < 1. \tag{4.24}$$

Remark 1. A physical interpretation of the sufficient condition (C) is as follows. Assuming the path-loss fading model $\mathbf{h}_{mq_i} = \bar{\mathbf{h}}_{mq_i} d_{mq_i}^{-\beta}$, where $\bar{\mathbf{h}}_{mq_i}$ contains normalized i.i.d. $\mathscr{CN}(0, 1)$ channel gains, d_{mq_i} is the distance between BS-m to user-i of cell-q, and β is the path-loss exponent. When the distance d_{mq_i} increases, the cross channel gains $\mathbf{h}_{mq_i}$, $m \neq q$ are smaller, which shrinks the elements in the cross interference matrix $\mathbf{G}_{mq}$. Thus, the positive off-diagonal elements of $\mathbf{S}$, which are the Frobenius norm of $\mathbf{G}_{qq}^{-1}\mathbf{G}_{mq}$'s, also become smaller. This results in a smaller spectral radius of $\mathbf{S}$. Thus, the more apart a MS from the BSs of other cells, the higher chance of $\varrho(\mathbf{S})$ being less than 1, which then guarantees the existence and uniqueness of the NE.

It is to be noted that while the sufficient condition (C) is obtained from the contraction mapping property of the power update, the direct characterization of the NE can be utilized to draw the necessary condition of the NE existence.

4.2.3.2 The Necessary Condition

We next examine the necessary condition for the existence and uniqueness of the NE of game $\mathscr{G}$. Before proceeding, some definitions of related mathematical terms to be used in the derivations are presented as follows.

Definition 4.1 ([2, 4]). A square matrix $\mathbf{A}$ is a $\mathbf{Z}$-matrix if all of its off-diagonal elements are non-positive. A square matrix $\mathbf{A}$ is a $\mathbf{P}$-matrix if all of its principal minors are positive. A square matrix that is both a $\mathbf{Z}$-matrix and a $\mathbf{P}$-matrix is called an $\mathbf{M}$-matrix.

Besides the above definition, there are several equivalent characterizations of an $\mathbf{M}$-matrix [2]. It is to be noted that if a matrix is an $\mathbf{M}$-matrix, it is invertible and the inverse is a positive matrix [2].

Proposition 4.6. *The NE of game $\mathscr{G}$ exists if and only if the following matrix*

$$(C1): \quad \mathbf{G} = \begin{bmatrix} \mathbf{G}_{11} & -\mathbf{G}_{21} & \cdots & -\mathbf{G}_{Q1} \\ -\mathbf{G}_{12} & \mathbf{G}_{22} & \cdots & -\mathbf{G}_{Q2} \\ \vdots & \vdots & \ddots & \vdots \\ -\mathbf{G}_{1Q} & -\mathbf{G}_{2Q} & \cdots & \mathbf{G}_{QQ} \end{bmatrix} \quad (4.25)$$

is an $\mathbf{M}$-matrix.

Proof. First, if the NE of game $\mathscr{G}$ exists, then an intersection point of the BR functions must exist. At the NE, (4.15) is equivalent to

$$\mathbf{G}_{qq}\mathbf{p}_q^\star - \sum_{m \neq q}^{Q} \mathbf{G}_{mq}\mathbf{p}_m^\star = \mathbf{1}\sigma^2, \ \forall q.$$

Reorganizing the above set of equations into a matrix form, one has

$$\mathbf{G}\mathbf{p}^\star = \mathbf{1}\sigma^2,$$

where $\mathbf{G}$ is previously defined. Note that $\mathbf{G}$ is a $\mathbf{Z}$-matrix, as its off-diagonal elements are all non-positive. Since there exists $\mathbf{p}^\star > 0$ to make $\mathbf{G}\mathbf{p}^\star = \mathbf{1}\sigma^2 > 0$, this implies $\mathbf{G}$ being an $\mathbf{M}$-matrix by its characterization (Condition I_{28}, Theorem 6.2.3 in [2]).

Conversely, if $\mathbf{G}$ is an $\mathbf{M}$-matrix, its inverse exists and is a positive matrix [2]. Thus, there exists a vector $\mathbf{p}^\star = \mathbf{G}^{-1}\mathbf{1}\sigma^2 > 0$, and $\mathbf{p}^\star$ satisfies the condition of being an intersection point of the BR functions in (4.15). As a result, an NE must exist.

As previously mentioned, there are various characterizations of an **M**-matrix [2] that one can utilize to verify whether matrix **G** is one of the type.

4.3 Comparison to the Fully Coordinated Design

In Sect. 4.2, a fully decentralized approach in the multi-cell downlink design was investigated and the NE of the multicell precoding game was established. However, it is well-known that the NE need not be Pareto-efficient [6]. Via the coordination among the BSs, significant power reduction can be obtained by jointly designing all the beamformers at the same time. Nonetheless, this advantage may come with the cost of message passing among the BSs as explained later in this section. This section is to review such a fully coordinated multicell downlink beamforming system [5], where the weighted total transmit power of all the cells is jointly minimized. A comparison between the two designs is presented in the end of the section.

Let $\mathbf{u}_{q_i}$ be the beamforming vector for user-i of cell-q with the coordinated design, and let $\mathbf{U}_q = [\mathbf{u}_{q_1}, \ldots, \mathbf{u}_{q_K}]$. The problem of jointly minimizing the weighted total transmit power of the Q cells is stated as follows

$$\underset{\mathbf{U}_1,\ldots,\mathbf{U}_Q}{\text{minimize}} \quad \sum_{q=1}^{Q} \omega_q \|\mathbf{U}_q\|_F^2 \tag{4.26}$$

$$\text{subject to} \quad \frac{|\mathbf{u}_{q_i}^H \mathbf{h}_{qq_i}|^2}{\sum_{j\neq i}^{K} |\mathbf{u}_{q_j}^H \mathbf{h}_{qq_i}|^2 + \sum_{m\neq q}^{Q} \sum_{j=1}^{K} |\mathbf{u}_{m_j}^H \mathbf{h}_{mq_i}|^2 + \sigma^2} \geq \gamma_{q_i},$$

where ω_q is the weight factor at BS-q, and $\sum_{q=1}^{Q} \omega_q = 1$. For a given $\boldsymbol{\omega} = [\omega_1, \ldots, \omega_Q] \geq 0$, the optimal solution to (4.26) represents an optimal trade-off point between the cells' power consumptions. Certainly, this point is Pareto-optimal, i.e., one cannot further reduce the power consumption at one cell without increasing the power consumption at (one or more) other cells.

The optimization problem (4.26) is convex, since the objective function is convex and the SINR constraints can be transformed into convex second order conic (SOC) constraints [5]. Thus, its optimal solution can be obtained using any conic solution package or standard convex optimization algorithm. This approach, however, is fully centralized. On the other hand, by exploiting the dual problem, this problem can be solved in a distributed fashion with message passing among the BSs. Via the Lagrangian technique, the dual problem of (4.26) is equivalent to the virtual dual uplink problem [5]

$$\underset{\{v_{q_i}\},\{\hat{\mathbf{u}}_{q_i}\}}{\text{minimize}} \sum_{q=1}^{Q}\sum_{i=1}^{K} v_{q_i}\sigma^2 \tag{4.27}$$

$$\text{subject to } \frac{v_{q_i}\left|\hat{\mathbf{u}}_{q_i}^H\mathbf{h}_{qq_i}\right|^2}{\displaystyle\sum_{m=1}^{Q}\sum_{j=1}^{K} v_{m_j}\left|\hat{\mathbf{u}}_{q_i}^H\mathbf{h}_{qm_j}\right| + \omega_q\hat{\mathbf{u}}_{q_i}^H\hat{\mathbf{u}}_{q_i}} \geq \frac{\gamma_{q_i}}{1+\gamma_{q_i}},$$

where the optimization is taken over the sum transmit power of the uplink powers v_{q_i}'s and the receive beamformer vectors $\hat{\mathbf{u}}_{q_i}$'s. Note that the optimal solution of the uplink powers v_{q_i}'s can be obtained as a unique fixed point of the following iteration [5]

$$v_{q_i}^{(n+1)} = \frac{\gamma_{q_i}}{1+\gamma_{q_i}} \times \frac{1}{\mathbf{h}_{qq_i}^H\left(\Sigma_q(\{v_{m_j}^{(n)}\})\right)^{-1}\mathbf{h}_{qq_i}} \tag{4.28}$$

with

$$\Sigma_q\left(\{v_{m_j}^{(n)}\}\right) = \sum_{m=1}^{Q}\sum_{j=1}^{K} v_{m_j}^{(n)}\mathbf{h}_{qm_j}\mathbf{h}_{qm_j}^H + \omega_q\mathbf{I}.$$

This iteration function is shown to be *standard* [5], which guarantees its convergence to a unique solution, if the problem is feasible [19]. It is to be noted that the feasibility study of this coordinated design has not been done in [5]. However, if the NE of the competitive design exists, the coordinated design must be feasible. Similar to the single-cell problem, given the optimal uplink transmit power $v_{q_i}^\star$, the optimal receive beamformers $\hat{\mathbf{u}}_{q_i}$ is the MMSE receiver, i.e.,

$$\hat{\mathbf{u}}_{q_i} = \left(\Sigma_q(\{v_{m_j}^\star\})\right)^{-1}\mathbf{h}_{qq_i}. \tag{4.29}$$

In addition, the optimal beamformer of the coordinated design, $\mathbf{u}_{q_i}$, can be found as a scaled version of $\hat{\mathbf{u}}_{q_i}$ by a factor $\sqrt{\delta_{q_i}}$ [5], i.e., $\mathbf{u}_{q_i} = \sqrt{\delta_{q_i}}\hat{\mathbf{u}}_{q_i}$. As all the SINR constraints in (4.26) are met with equality at optimality, substituting $\mathbf{u}_{q_i} = \sqrt{\delta_{q_i}}\hat{\mathbf{u}}_{q_i}$, the KQ SINR constraints can be written as

$$\delta_{q_i}\frac{\left|\hat{\mathbf{u}}_{q_i}^H\mathbf{h}_{qq_i}\right|^2}{\gamma_{q_i}} - \sum_{j\neq i}^{K}\delta_{q_j}\left|\hat{\mathbf{u}}_{q_j}^H\mathbf{h}_{qq_i}\right|^2 - \sum_{m\neq q}^{Q}\sum_{j=1}^{K}\delta_{m_j}\left|\hat{\mathbf{u}}_{m_j}^H\mathbf{h}_{mq_i}\right|^2 = \sigma^2. \tag{4.30}$$

In order to find δ_{q_i}, one needs to solve this set of KQ equations. Define a matrix $\mathbf{F}$ of size $KQ \times KQ$, and its components as

$$[\mathbf{F}]_{K(q-1)+i, K(m-1)+j} = \begin{cases} \dfrac{\left|\hat{\mathbf{u}}_{q_i}^H \mathbf{h}_{qq_i}\right|^2}{\gamma_{q_i}} & , \text{if } q = m, i = j \\ -\left|\hat{\mathbf{u}}_{m_j}^H \mathbf{h}_{mq_i}\right|^2 & , \text{otherwise} \end{cases}$$

with $i, j = 1, \ldots, K$, and $q, m = 1, \ldots, Q$. Then, one has

$$\boldsymbol{\delta} = [\delta_{1_1}, \ldots, \delta_{1_K}, \ldots, \delta_{Q_1}, \ldots, \delta_{Q_K}]^T = \mathbf{F}^{-1} \mathbf{1} \sigma^2. \tag{4.31}$$

In summary, with the fully coordinated multicell system, the following three-step algorithm is needed to find the jointly optimal beamformers [5]:

(i) Apply the iteration (4.28) to find the fixed point $v_{q_i}^\star$.
(ii) Find the receive beamformer $\hat{\mathbf{u}}_{q_i}$ for the dual uplink channel as in (4.29).
(iii) Find the scaling factor δ_{q_i}.

In [5], the authors argued that the above algorithm can be implemented in a distributed manner under the condition of channel reciprocity. More specifically, when the uplink and downlink channels are reciprocal of each other, the virtual dual uplink is the real uplink. Thus, the iteration (4.28) in step (i) can be performed locally at each BS with local information. In particular, at BS-q, $\mathbf{h}_{qq_i}$ is typically known and Σ_q is effectively the covariance matrix of the received signal in the uplink direction. In step (ii), the receive beamformer $\hat{\mathbf{u}}_{q_i}$ can be easily obtained. Finally, step (iii) can be implemented iteratively where each δ_{q_i} is determined locally to meet its corresponding SINR target (assuming all other δ_{q_i}'s are fixed) until convergence. Overall, the distributed implementation of the coordinated design requires channel reciprocity and the synchronization among the BSs to be in the uplink phase or the downlink phase together [5]. It is worth mentioning that in practice the uplink and downlink channels usually operate in separate frequency bands in the frequency division duplexing (FDD) mode. Channel reciprocity is therefore hard to realize.

Obviously, if the condition on channel reciprocity is not true, each BS in the coordinated design needs to know all the channels from itself to all the MSs in the system. A message passing scheme among the cells is then required to jointly update the dual variables v_{q_i}'s. In addition, certain synchronization among the BSs is desired, i.e., step (iii). These are the main differences to the competitive design considered previously in this chapter, where the beamforming design is performed locally at each cell without any message exchanges and synchronization. These differences prompt us to investigate a new game that retains the advantages of the power minimization game $\mathscr{G}$, i.e., fully distributed implementation, no message passing, and no synchronization, and possibly approaches the performance established by the coordinated design. This concern will be addressed in the next section.

4.4 Multicell Downlink Beamforming Game with Pricing

4.4.1 Problem Formulation

This section begins with a numerical example of the power consumption at a two-cell system with both the coordinated and competitive designs. Considered is a system with two cells and two MSs per cell with target SINR $\gamma_{q_i} = 10$ (10dB). It is assumed that each BS is equipped with 3 antennas and the distance between the two BSs is normalized to 2. Each MS is located between the two BSs at a distance $d = 0.6$ from its connected BS. All the intra-cell and inter-cell channel coefficients are generated from i.i.d. Gaussian random variables, using the path-loss model with the path-loss exponent of 3. The background noise power σ^2 is 0.01. Figure 4.2 displays the power consumption at the two BS with the competitive design, i.e., the NE point of game $\mathscr{G}$, relatively to the coordinated design. It should be noted that the power consumption at the cell is lower-bounded by the minimum power requirement to meet its users' SINR target, in the absence of ICI. It can be drawn from the figure that the NE point of the competitive design is relatively inefficient, compared to the Pareto-optimal curve established by the coordinated design.

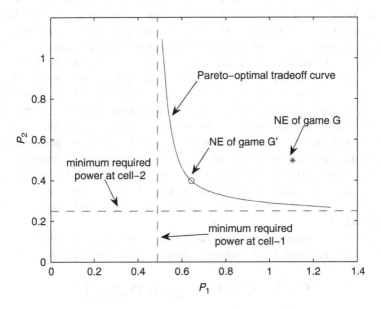

Fig. 4.2 Power consumption in two cells: competitive design vs. coordinated design

An interesting question here is whether one can modify the utility function at each player such that the game becomes more cooperative and its equilibria possibly lie on the boundary of the Pareto-optimal trade-off surface, e.g., point "o" in Fig. 4.2. In this section, a new game with pricing consideration, namely game $\mathscr{G}'$ is to be examined. By introducing a pricing component to each player's utility function, the players now voluntarily cooperate with others by minimizing their inducing interference to the others as well as minimizing their own transmit power at the same time. In fact, it shall be shown that the point "∘" can be obtained at the NE of the modified game $\mathscr{G}'$.

Now, suppose that BS-q has additional information about the channel to the users in other cells, and it performs the following optimization

$$\underset{\mathbf{V}_q}{\text{minimize}} \quad \sum_{i=1}^{K} \left\| \mathbf{v}_{q_i} \right\|^2 + \sum_{m \neq q}^{Q} \sum_{j=1}^{K} \pi_{qm_j} \left\| \mathbf{V}_q^H \mathbf{h}_{qm_j} \right\|^2 \qquad (4.32)$$

$$\text{subject to} \quad \frac{\left| \mathbf{v}_{q_i}^H \mathbf{h}_{qq_i} \right|^2}{\sum_{j \neq i}^{K} \left| \mathbf{v}_{q_j}^H \mathbf{h}_{qq_i} \right|^2 + r_{-q_i}} \geq \gamma_{q_i}, \ \forall i,$$

where $\pi_{qm_j} \geq 0$ is the pricing factor and $\left\| \mathbf{V}_q^H \mathbf{h}_{qm_j} \right\|^2$ is the interference at user-j of cell-m, caused by BS-q.[3]

It is to be noted that unlike the coordinated design in Sect. 4.3, this downlink beamforming game can be implemented at the system where partial information is available. More specifically, if the channel to user-i at cell-m is known at BS-q, a pricing factor $\pi_{qm_j} > 0$ is set to motivate cell-q to adopt a more sociable strategy by steering its beamformers to the directions that cause less interference (damage) to other cells. Otherwise, the pricing factor π_{qm_j} is set to 0. Certainly, the nature of the game being played among the players is no longer purely competitive. Through pricing, it is possible to improve system performance by inducing cooperation among the players, and yet maintaining the decentralized nature of the game. In general, the pricing factors π_{qm_j} should be tuned such that a largest possible enhancement in the overall system is obtained [13]. In a dynamic pricing scheme, the pricing factors can be jointly decided and constantly exchanged among the BSs. However, in order to reduce the system overhead, it is assumed that the prices are chosen a priori and remain fixed during the game being played. This assumption may be motivated by a system with a system designer, who informs the prices to the players in advance.

The game with pricing consideration is practically the same game as $\mathscr{G}$ with different payoff function. Mathematically, the new game is defined as

$$\mathscr{G}' = \left(\Omega, \left\{ \mathscr{P}_q(\mathbf{V}_{-q}) \right\}_{q \in \Omega}, \left\{ s_q(\mathbf{V}_q) \right\}_{q \in \Omega} \right),$$

---

[3]To avoid any confusion with the notation $\mathbf{w}_{q_i}$ used in Sect. 4.2 and $\mathbf{u}_{q_i}$ used in Sect. 4.3, $\mathbf{v}_{q_i}$ is denoted as the beamformer for user-i of cell-q in the competitive design with pricing consideration within this section. Similarly, $\mathbf{V}_q$ is used instead of $\mathbf{W}_q$ and $\mathbf{U}_q$. Likewise, let $\mathbf{q}_q \in \mathbb{R}^K$ denote the allocated power vector for the K users in cell-q, instead of $\mathbf{p}_q$ in Sect. 4.2.

where $s_q(\mathbf{V}_q) = t_q(\mathbf{V}_q) + \sum_{m \neq q}^{Q} \sum_{j=1}^{K} \pi_{qm_j} \|\mathbf{V}_q^H \mathbf{h}_{qm_j}\|^2$ is the utility function at player-q. Our interest in this part is to study whether game $\mathscr{G}'$ eventually converges to an NE and whether the NE is unique. A feasible strategy profile $\{\mathbf{V}_q^\star\}_{q=1}^{Q}$ is an NE of game $\mathscr{G}'$ if

$$s_q(\mathbf{V}_q^\star) \leq s_q(\mathbf{V}_q), \quad \forall \mathbf{V}_q \in \mathscr{P}_q(\mathbf{V}_{-q}^\star), \quad \forall q \in \Omega. \tag{4.33}$$

4.4.2 NE Existence and Uniqueness

This section studies the existence and uniqueness of the NE of the new game $\mathscr{G}'$. First of all, given the strategy of other player $\mathbf{V}_{-q}$, the optimal strategy for player-q can be obtained by solving the optimization problem (4.32). Note that problem (4.32) is convex, as the constraints are SOC and the objective function is quadratic. This useful observation enables us to find its optimal solution via convex optimization. In addition, uplink-downlink duality can be exploited to devise the optimal solution for this problem. The following theorem establishes the analytical steps to find the solution.

Theorem 4.1. *The optimal transmit beamforming problem* (4.32) *can be solved via a dual virtual uplink channel*

$$\underset{\substack{\mu_{q_1}, \dots, \mu_{q_K} \\ \hat{\mathbf{v}}_{q_1}, \dots, \hat{\mathbf{v}}_{q_K}}}{\text{minimize}} \sum_{i=1}^{K} \mu_{q_i} r_{-q_i} \tag{4.34}$$

$$\text{subject to } \frac{\mu_{q_i} |\hat{\mathbf{v}}_{q_i}^H \mathbf{h}_{qq_i}|^2}{\sum_{j \neq i}^{K} \mu_{q_j} |\hat{\mathbf{v}}_{q_i}^H \mathbf{h}_{qq_j}|^2 + \hat{\mathbf{v}}_{q_i}^H \Phi_q(\{\pi_{qm_j}\}) \hat{\mathbf{v}}_{q_i}} \geq \gamma_{q_i}, \forall i,$$

where $\Phi_q(\{\pi_{qm_j}\}) = \sum_{m \neq q}^{Q} \sum_{j=1}^{K} \pi_{qm_j} \mathbf{h}_{qm_j} \mathbf{h}_{qm_j}^H + \mathbf{I}$ *is treated as the noise covariance matrix at the BS, and the optimization is taken over the weighted sum-power of the uplink power* μ_{q_i} *and the receive beamformer vectors* $\hat{\mathbf{v}}_{q_i}$. *The optimal* $\mathbf{v}_{q_i}$ *is a scaled version of the optimal* $\hat{\mathbf{v}}_{q_i}$.

Proof. The proof of this theorem can be found in [10]. □

To find the optimal solution of μ_{q_i} to problem (4.34), the following fixed point iteration can be utilized

$$\mu_{q_i}^{(n+1)} = \frac{\gamma_{q_i}}{1+\gamma_{q_i}} \times \frac{1}{\mathbf{h}_{qq_i}^H \left(\sum_{j=1}^{K} \mu_{q_j}^{(n)} \mathbf{h}_{qq_j} \mathbf{h}_{qq_j}^H + \Phi_q(\{\pi_{qm_j}\}) \right)^{-1} \mathbf{h}_{qq_i}}. \quad (4.35)$$

Using the *standard* function property, this iteration is guaranteed to converge to a unique fixed-point if the primal problem (4.32) is feasible.[4]

Having solved problem (4.32), it is clear that its solution resembles the solution of the typical downlink beamforming problem (4.4). Thus, many properties of problem (4.4)'s solution, i.e., Propositions 4.1–4.4, also hold. This observation is summarized in the following lemma.

Lemma 4.2. *Given fixed pricing factors π_{qm_j}'s, Propositions 4.1–4.4 associated with the solution of problem (4.4) are also applicable to the solution of problem (4.32).*

Proof. Proposition 4.1 is straightforward. As the feasibility depends only on the constraints, if problem (4.4) is feasible, problem (4.32) is also feasible. Thus, Proposition 4.2 also holds. Proposition 4.3 comes directly from the fact that the solution $\mathbf{v}_{q_i}$ of problem (4.4) is a scaled version of $\hat{\mathbf{v}}_{q_i}$. Proposition 4.4 also holds, following the same procedure given in Sect. 4.2.2. That is, given $\mathbf{r}_{-q}$ as the total interference induced by Ω_{-q} plus background noise and $\tilde{\mathbf{v}}_{q_i}$ as the beam pattern corresponding to π_{qm_j}, the optimal allocated power vector $\mathbf{q}_q \in \mathbb{R}^K$ for the K users at BS-q is $\mathbf{q}_q = \mathbf{K}^{-1}\mathbf{r}_{-q}$, where $\mathbf{K} \in \mathbb{R}^{K \times K}$ is defined as $[\mathbf{K}]_{i,i} = (1/\gamma_{q_i})|\tilde{\mathbf{v}}_{q_i}^H \mathbf{h}_{qq_i}|^2$ and $[\mathbf{K}]_{i,j} = -|\tilde{\mathbf{v}}_{q_j}^H \mathbf{h}_{qq_i}|^2$ if $i \neq j$.

Denote $\mathbf{K}_{mq} \in \mathbb{R}^{K \times K}, m \neq q$ as the ICI matrix, where $[\mathbf{K}_{mq}]_{i,j} = |\tilde{\mathbf{v}}_{m_j}^H \mathbf{h}_{mq_i}|^2$. Then, one has $\mathbf{r}_{-q} = \sum_{m \neq q}^{Q} \mathbf{K}_{mq}\mathbf{q}_m + \mathbf{1}\sigma^2$. From Lemma 4.2, subject to the strategy of Ω_{-q}, the best response strategy of the player-q with pricing consideration is

$$\mathbf{q}_q^\star = \mathrm{BR}_q'(\mathbf{q}_{-q}) = \mathbf{K}_q^{-1}\left(\sum_{m \neq q}^{Q} \mathbf{K}_{mq}\mathbf{q}_m + \mathbf{1}\sigma^2 \right). \quad (4.36)$$

Lemma 4.3. *With pricing consideration, the best response function of player-q is standard.*

Proof. The proof is the same as that in Lemma 4.1.

Since the best response function $\mathrm{BR}_q'(\mathbf{q}_{-q})$ is *standard*, from any starting point $\mathbf{q}^{(0)}$, the iteration $\mathbf{q}_q^{(t+1)} = \mathrm{BR}_q'(\mathbf{q}_{-q}^{(t)})$ will surely converge to a fixed point (if it exists), which is the NE of game $\mathscr{G}'$. The necessary condition for the existence of the NE in game $\mathscr{G}'$ is similar to that of game $\mathscr{G}$, established in Proposition 4.6, i.e., the matrix $\mathbf{K} \in \mathbb{R}^{KQ \times KQ}$, in the same form as $\mathbf{G}$ in (4.25) with $\mathbf{K}_{qq}$ and $\mathbf{K}_{mq}$ replacing

---

[4]The proof for this function to be *standard* is similar to the ones in [18].

$\mathbf{G}_{qq}$ and $\mathbf{G}_{mq}$, is an **M**-matrix. Likewise, thanks to Proposition 4.5, if $\varrho(\mathbf{S}') < 1$, where $\mathbf{S}' \in \mathbb{R}^{K \times K}$ is defined as $[\mathbf{S}']_{qq} = 0$ and $[\mathbf{S}']_{q,m} = \left\| \mathbf{K}_{qq}^{-1} \mathbf{K}_{mq} \right\|_F$ if $m \neq q$, game $\mathscr{G}'$ is also guaranteed to admit a unique NE.

To this point, one may wonder how efficient the NE of game $\mathscr{G}'$ is compared to the NE of game $\mathscr{G}$ and the Pareto-optimal trade-off curve. Although this chapter does not present a concrete proof to the proposition that the NE of game $\mathscr{G}'$ is more efficient than that of game $\mathscr{G}$, all simulations show that with right pricing factors, this proposition is true. In fact, with a certain pricing scheme deployed at all the BSs, the NE of game $\mathscr{G}'$ is able to approach the Pareto-optimal trade-off curve. The characterization of this pricing scheme is given in the following.

Theorem 4.2. *Given the weight vector $\boldsymbol{\omega}$ for the coordinated design, and suppose that $v_{q_i}^\star$'s are the dual variables that satisfy the iteration* (4.28), *if the weight factors for game $\mathscr{G}'$ are set as*

$$\pi_{qm_j} = \frac{v_{m_j}^\star}{\omega_q}, \quad \forall i, \forall q, \tag{4.37}$$

the NE of game $\mathscr{G}'$ is exactly the solution of the coordinated design. That is, the NE lies on the Pareto-optimal trade-off curve.

Proof. When the weight factors for game $\mathscr{G}'$ are set as in (4.37), the fixed-point iteration (4.35) becomes

$$\omega_q \mu_{q_i}^{(n+1)} = \frac{\gamma_{q_i}}{1 + \gamma_{q_i}} \times \frac{1}{\mathbf{h}_{qq_i}^H \left(\sum_{j=1}^{K} \omega_q \mu_{q_j}^{(n)} \mathbf{h}_{qq_j} \mathbf{h}_{qq_j}^H + \omega_q \Phi_q(\{v_{m_j}^\star / \omega_q\}) \right)^{-1} \mathbf{h}_{qq_i}}. \tag{4.38}$$

Comparing to fixed-point iteration (4.28), it is obvious that the unique fixed-point of the above iteration satisfies $\omega_q \mu_{q_i}^\star = v_{q_i}^\star$. As a result, $\hat{\mathbf{v}}_{q_i} = \omega_q \hat{\mathbf{u}}_{q_i}$. That is, the beam pattern set for the users of game $\mathscr{G}'$ is the same as the beam pattern set in the coordinated design. Thus, it is left to determine that the beamformers of the two designs are indeed the same. Note that the coordinated design determines the scaling factor δ_{q_i} to $\hat{\mathbf{u}}_{q_i}$ by either using matrix inversion, c.f. Eq. (4.31), or each MS sets a δ_{q_i} to meet its corresponding SINR constraint assuming all other δ_{q_i}'s are fixed [5]. The convergence of the second method can be proved by the *standard* function technique [5], which effectively explains the convergence to a unique fixed-point. On the other hand, at each time instance, BS-q in game $\mathscr{G}'$ determines the allocated powers (equivalently the scaling factors to $\hat{\mathbf{v}}_{q_i}$) to satisfy the SINR constraints at its users. Due to the uniqueness of the fixed-point, the game played in $\mathscr{G}'$ has to converge to the same solution as the coordinated design.

Theorem 4.2 is significant in the sense that the fully coordinated design can still be interpreted as a competitive game with the *right* pricing scheme. Our next task is to study how to implement such a pricing scheme, under two game scenarios: (a) game with complete information and (b) game with incomplete information.

In a game with complete information, it is assumed that the system designer knows all the game parameters, including the channels and the QoS requirements. Each BS is also assumed to fully know its channels to all the MSs. The designer then can exactly decide the optimal dual variables $v_{q_i}^\star$'s, which are used to determine the optimal prices in (4.37). As stated in Theorem 4.2, the game with pricing consideration will be Pareto-optimal. Interestingly, it has been recently shown in [1] that pricing can allow the system designer to locate the NE point to any feasible point in a broad class of power allocation games under complete information. Our result given in Theorem 4.2 has established a similar result for the beamforming and power allocation game in a multicell system.

In a game with incomplete information, it is assumed that neither the BSs nor the system designer fully know all the channels. Thus, implementing the pricing scheme (4.37) is no longer possible. In this case, the designer may help the BSs to search for good pricing factors. Here, the same mechanism exploited in [13] is employed due to its simplicity. More specifically, the designer lets the BSs play game $\mathscr{G}$ (no pricing) and obtain the NE. Then, each BS sets its pricing factors π_{qm_j} to a same value $c > 0$ (initially large), which is informed by the designer, and game $\mathscr{G}'$ is played between the BSs. After dividing the pricing factor c by a positive factor of Δc, game $\mathscr{G}'$ is played again and its NE is re-measured. The procedure is repeated if the sum of the utility functions at the new equilibrium is smaller than that of the previous instance. Otherwise, the procedure is stopped and all the pricing factors are set to the same factor, called c_{BEST}. As shall be shown in the simulation, this technique performs very well in improving the NE's efficiency.

4.5 Numerical Results

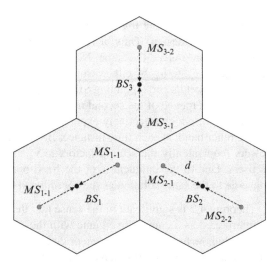

Fig. 4.3 A multicell system configuration with 3 cells, 2 users per cell. Of the two users at each cell, one stays close to the borders with other cells, one is far away

This section presents some numerical results to validate our findings. In particular, the feasibility of the coordinated design and the probability of existence of an NE of games $\mathscr{G}$ and $\mathscr{G}'$ are compared. Also compared are the average total transmit powers (of all the cells) of the three designs. Consider a multicell network as illustrated in Fig. 4.3, composed of 3 cells with 2 users per cell. It is assumed that the BSs are equidistant, and their distance is normalized to 2. The distance between a MS and its serving BS is also set the same, at d. Of the two MSs at each cell, one is located close to the borders with other cells, whereas the second one is far away. The same target SINRs are set at each MS, either $\gamma_{q_i} = 0$ dB or $\gamma_{q_i} = 10$ dB. The AWGN power spectral density σ^2 is set at 0.01. The channels from a BS to a MS are generated from i.i.d. Gaussian random variables using the path-loss model with the path-loss exponent of $\beta = 3$ and the reference distance of 1 corresponding to MSs at the cell-edge. As the distance d is varied, 10,000 channel realizations at each value d are used to plot the probability of the existence of a stable operating point in Fig. 4.4 and the average total transmit power in Fig. 4.5.

In the competitive design with pricing consideration, it is assumed that each BS also knows the channels to the MSs that are close to the cell boundary at the other two cells. For example, base-station BS-1 knows its channels to MS-2_1 and MS-3_1. Certainly, these two MSs are subject to a much higher ICI level from cell-1 than the others. Each BS then takes advantage of this extra information to improve efficiency of the NE of game $\mathscr{G}'$ with pricing. The pricing scheme for games with incomplete information as discussed in Sect. 4.4 is applied in this simulation.

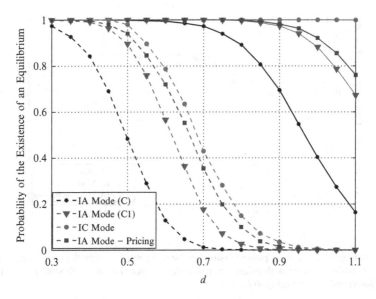

Fig. 4.4 Probability of existence of a stable operating point versus d by evaluating conditions (C) and (C1) and numerically examining the convergence of game $\mathscr{G}$, the coordinated design, and game $\mathscr{G}'$ to meet the target SINRs: $\gamma_{q_i} = 10$ dB (*dashed lines*) and $\gamma_{q_i} = 0$ dB (*solid lines*)

Figure 4.4 displays the probability of existence of a stable operating point as the function of the MS-BS distance d by evaluating whether condition (C) is satisfied and numerically examining the convergence of game $\mathscr{G}$ (which matches with condition (C1)), the coordinated design, and game $\mathscr{G}'$. From the figure, as the MSs get closer to their BS, a higher probability of the existence of a stable operating point for all three designs is observed. This is due to the fact that the stronger intra-cell channels allow a BS to transmit at a lower power level to meet its target SINRs, which then causes lower ICI. With the competitive design, a lower level of ICI certainly guarantees a higher probability of existence of an NE. On the other hand, with the pricing consideration, the whole system may further reduce the ICI. As a result, the existence probability of game $\mathscr{G}'$ is higher than that of game $\mathscr{G}$. Finally, with the coordinated design, where the ICI is fully managed, the feasibility of finding a solution that meets all the target SINRs is certainly higher than finding one in both games $\mathscr{G}$ and $\mathscr{G}'$. Note that conditions (C) and (C1) can be easily examined to verify the existence of the NE of the competitive design. In case of not meeting conditions (C) or (C1), one may attempt to switch the network into the design with pricing consideration or the fully coordinated design to improve the convergence probability of the whole system. Figure 4.4 also shows that a lower target SINR, which requires lower transmit power (lower ICI), would induce a higher chance of finding the solutions in all three designs.

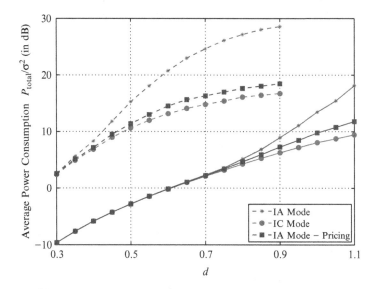

Fig. 4.5 Average total transmit power versus d of the competitive design, the coordinated design, and the competitive design with pricing consideration to meet the target SINRs: $\gamma_{q_i} = 10$ dB (*dashed lines*) and $\gamma_{q_i} = 0$ dB (*solid lines*)

Of all cases where game $\mathscr{G}$ converges, the total transmit power P_{total} at the 3 BSs with three designs are averaged and compared in Fig. 4.5 in the form of $P_{\text{total}}/\sigma^2$. Note that the weight factors ω_q's of the coordinated design are set equal to each other to minimize the design's sum transmit power. At small inter-cell MS-BS distances, it can be seen that the power usages of all the designs are very low and their difference is rather marginal. Again, this is due to the fact that the intra-cell channels are strong and the ICI is too small. However, as d increases, the effect of ICI becomes significant. Since the competitive design does not attempt to control the ICI, its NE point becomes inefficient compared to the Pareto-optimal frontier established by the coordinated design. On the other hand, should a BS know the channels to the MSs at other cells, it can alter its strategy by playing the game with pricing consideration $\mathscr{G}'$. In fact, using the aforementioned procedure to determine the pricing factor, the NE of game $\mathscr{G}'$ is almost Pareto-optimal. It is worth noting that this result is obtained even thought each BS does not possess full channel knowledge from itself to all the users.

To illustrate the convergence behaviors of the multicell downlink beamforming games $\mathscr{G}$ and $\mathscr{G}'$ and compare the transmit powers of the two designs, two examples are selected and displayed in Figs. 4.6 and 4.7. In both games, it is assumed that all the BSs perform simultaneous power update at each time instance. The transmit power of each cell is then displayed after each iteration. The system configuration is the one in Fig. 4.3, with $\gamma_{q_i} = 10$ dB and $d = 0.6$.

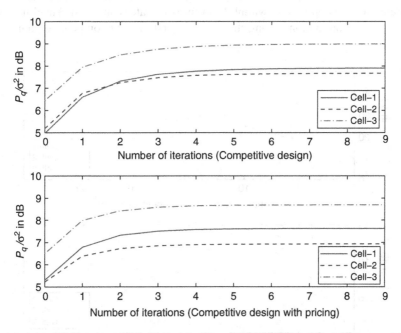

Fig. 4.6 A converging example of the downlink beamforming games $\mathscr{G}$ and $\mathscr{G}'$ in a multicell system: the sum power of each cell versus the number of iterations with $\varrho(\mathbf{S}) = 0.7332$, $\varrho(\mathbf{S}') = 0.6256$, and the corresponding matrices $\mathbf{G}$ and $\mathbf{K}$ are M-matrices

In the first example, $\varrho(\mathbf{S})$ and $\varrho(\mathbf{S'})$ are calculated at 0.7332 and 0.6256, respectively. The power updates of both games displayed in Fig. 4.6 clearly show the convergence of the two designs. Figure 4.6 also shows the benefit of using the design with pricing consideration, where the transmit power at each cell is reduced, compared to that of the purely competitive design. For this particular example, the price is set at 0.1388.

In the second example, $\varrho(\mathbf{S})$ and $\varrho(\mathbf{S'})$ are calculated at 2.0016 and 0.9815, respectively. In can be seen from Fig. 4.7 that the power updates of game $\mathscr{G}$ do not converge. Interestingly, with the price set at 0.551, the design with pricing consideration eventually converges. This behavior clearly indicates the benefit of adopting a more cooperative strategy at each cell by exploiting the extra channel information to other cells.

4.6 Concluding Remarks

This chapter has studied the problem of downlink beamformer designs in a multicell system via game theory. Given the QoS requirements at the users in its cell, each BS determines its optimal downlink beamformer strategy in a distributed manner, without any coordination among the cells. At first, a fully competitive game was

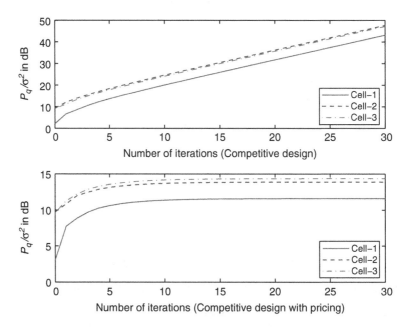

Fig. 4.7 An example of the downlink beamforming game that diverges in game $\mathscr{G}$ and but converges in game $\mathscr{G'}$: the sum power of each cell versus the number of iterations with $\varrho(\mathbf{S}) = 2.0016$, and $\varrho(\mathbf{S'}) = 0.9815$. In this case, the corresponding matrix $\mathbf{K}$ is an M-matrix, but $\mathbf{G}$ is not

considered, where each BS greedily minimizes its transmit power. The necessary and sufficient conditions guaranteeing the existence and uniqueness of the NE of the game were then presented. In addition, a comparison between the competitive and coordinated designs was also presented. Finally, to improve the efficiency of the competitive game, a new game with pricing consideration was studied. Through the pricing mechanism, each BS can steer its beamformers in a more cooperative way to reduce its induced ICI to other cells. If each BS knows its channels to all MSs in the network, the new game is capable of obtaining the Pareto-optimal performance as the coordinated design, while retaining the distributed nature of a multicell game. Interestingly, the multicell beamforming game with pricing consideration can also be implemented with partial inter-cell channel knowledge.

References

1. Alpcan, T., Pavel, L.: Nash equilibrium design and optimization. In: Proc. Int. Conf. on Game Theory for Networks (GameNets' 09), pp. 164–170. Istanbul, Turkey (2009)
2. Berman, A., Plemmons, R.J.: Nonnegative matrices in the mathematical sciences. Academic, New York (1979)
3. Bertsekas, D.P., Tsitsiklis, J.N.: Parallel and Distributed Computation: Numerical Methods. Prentice-Hall, New Jersey (1989)
4. Cottle, R.W., Pang, J.S., Stone, R.E.: The linear complementarity problem. Academic, Cambridge, UK (1992)
5. Dahrouj, H., Yu, W.: Coordinated beamforming for the multicell multi-antenna wireless system. IEEE Trans. Wireless Commun. 9(5), 1748–1759 (2010)
6. Dubey, P.: Inefficiency of Nash equilibria. Math. Oper. Res. 11(1), 1–8 (1986)
7. Horn, R.A., Johnson, C.R.: Matrix Analysis. Cambridge University Press, New York (1985)
8. Lasaulce, S., Debbah, M., Altman, E.: Methodologies for analyzing equilibria in wireless games. IEEE Signal Process. Magazine 26(5), 41–52 (2009)
9. Luo, Z.Q., Pang, J.S.: Analysis of iterative waterfilling algorithm for multiuser power control in digital subscriber lines. EURASIP Journal on Advances in Signal Processing 2006(1), 024,012 (2006). DOI 10.1155/ASP/2006/24012. URL http://asp.eurasipjournals.com/content/2006/1/024012
10. Nguyen, D.H.N., Le-Ngoc, T.: Multiuser downlink beamforming in multicell wireless systems: A game theoretical approach. IEEE Trans. Signal Process. 59(7), 3326–3338 (2011)
11. Pang, J.S., Scutari, G., Facchinei, F., Wang, C.: Distributed power allocation with rate constraints in Gaussian parallel interference channels. IEEE Trans. Inform. Theory 54(8), 3471–3489 (2008)
12. Ren, T., La, R.J.: Downlink beamforming algorithms with inter-cell interference in cellular networks. IEEE Trans. Wireless Commun. 5(10), 2814–2823 (2006)
13. Saraydar, C.U., Mandayam, N.B., Goodman, D.J.: Efficient power control via pricing in wireless data networks. IEEE Trans. Commun. 50(2), 291–303 (2002)
14. Scutari, G., Palomar, D.P., Barbarossa, S.: Optimal linear precoding strategies for wideband noncooperative systems based on game theory – Part I: Nash Equilibria. IEEE Trans. Signal Process. 56(3), 1230–1249 (2008)
15. Scutari, G., Palomar, D.P., Barbarossa, S.: Optimal linear precoding strategies for wideband noncooperative systems based on game theory – Part II: Algorithms. IEEE Trans. Signal Process. 56(3), 1250–1277 (2008)
16. Scutari, G., Palomar, D.P., Barbarossa, S.: The MIMO iterative waterfilling algorithm. IEEE Trans. Signal Process. 57(5), 1917–1935 (2009)

17. Shum, K., Leung, K.K., Sung, C.W.: Convergence of iterative waterfilling algorithm for Gaussian interference channels. IEEE J. Select. Areas in Commun. **25**(6), 1091–1100 (2007)
18. Wiesel, A., Eldar, Y.C., Shamai, S.: Linear precoding via conic optimization for fixed MIMO receivers. IEEE Trans. Signal Process. **54**(1), 161–176 (2006)
19. Yates, R.D.: A framework for uplink power control in cellular radio systems. IEEE J. Select. Areas in Commun. **13**(7), 1341–1347 (1995)
20. Yu, W., Ginis, G., Cioffi, J.M.: Distributed multiuser power control for digital subscriber lines. IEEE J. Select. Areas in Commun. **20**(5), 1105–1115 (2002)
21. Yu, W., Lan, T.: Transmitter optimization for the multi-antenna downlink with per-antenna power constraints. IEEE Trans. Signal Process. **55**(6), 2646–2660 (2007)

Chapter 5
Block-Diagonalization Precoding in Multiuser Multicell MIMO Systems

In Chap. 4, multicell precoding designs were investigated with the objective of minimizing the transmit power at the BSs subject to SINR constraints at the MSs. In this chapter, the multicell precoders are designed to maximize the achievable data-rate of the MSs with power constraints at the BSs. In a MIMO system, space-division multiple-access (SDMA) can be applied at the BS to concurrently multiplex data streams for multiple MSs. With appropriate downlink precoding techniques at the BS, SDMA can significantly improve the system's spectral efficiency. Downlink precoding for a MIMO system has been an active area of research for many years. Dirty-paper coding (DPC) [4, 6, 21, 25] has been proved to be the capacity-achieving multiuser precoding strategy. However, due to its high complexity implementation that involves random nonlinear encoding and decoding, DPC can only serves a theoretical benchmark. Consequently, linear precoding techniques, such as zero-forcing (ZF) and block-diagonalization (BD) [5, 15, 20, 22], become appealing alternatives due to their simplicity and good performance. With BD precoding, the transmitted signal from the BS intended for a particular MS is restricted to be in the null space created by the downlink channels associated with all the other MSs. Therefore, all inter-user interference at the MSs can be fully suppressed.

This chapter investigates the multicell system where BD precoding is applied on a per-cell basis. Specifically, we consider BD precoding for the multicell system under the two operating modes: interference aware (IA) and interference coordination (IC) [11]. Under these two operating modes, each BS is required to transmit information data *only* to the MSs within its cell limits. The reason for considering BD precoding on a per-cell basis is due to simple implementation and good performance at the high SNR region [20].

Under the IA mode, each interference-aware MS shall measure the level of ICI and feed back this information to its connected BS [11]. Given the strategies from other BSs reflected by the ICI, each BS selfishly adjusts its precoding strategy to maximize the sum-rate for its connected MSs. Naturally, the IA mode represents a strategic noncooperative game (SNG) with the BSs being the rational players.

The study of precoding design for the multicell system under the IA mode using game theory is plentiful in literature [10, 12, 14, 17, 18]. In general, these works focus on studying the existence and uniqueness of the game's Nash equilibrium (NE). Different from these works, the consideration of BD precoding in this chapter allows us to examine a multicell SNG under a more general setting where there are multiple MSs per cell and each MS is equipped with multiple antennas. In order to characterize the multicell BD precoding game, the best response strategy at the each BS is first presented in closed-form. Then, it is shown that this WF best response strategy can be interpreted as a projection onto a closed and convex set. This interpretation shall allow us to study the uniqueness of the game's NE later on.

Under the IC mode, while each BS only transmits data information to the MSs within its cell limits, the precoders from all BSs are jointly designed to fully control the ICI [11]. This chapter then examines the BD precoding strategy to jointly maximize the weighted sum-rate (WSR) of the multicell system under the IC mode. Since this WSR maximization problem is shown to be nonconvex, it is generally difficult and computationally complex to find its globally optimal solution. Thus, our focus is on proposing a low-complexity algorithm to approximate the nonconvex WSR maximization into a sequence of simpler convex problems, which can be solved separately at the corresponding BS. In particular, each BS attempts to optimize its BD precoder to maximize the sum-rate for its connected MSs while doing its best in limiting the ICI induced to other cells through an interference-penalty mechanism. Simulation results show a significant improvement in the network sum-rate by the IC mode over the IA mode at the high ICI region.

In the latter part of this chapter, the precoding design in a multiuser multicell system is investigate with Block-Diagonalization—Dirty-Paper Coding (BD-DPC) being utilized in a per-cell basis. In BD-DPC, the information signals sent to the multiple users are encoded in sequence such that the receiver at any user does not see any inter-user interference due to the use of BD and DPC at the BS [4]. Thus, BD-DPC can take advantage of DPC to enhance the performance of BD precoding. Under the IA mode, this chapter attempts to characterize the NE of the BD-DPC multicell precoding game by examining the conditions for its existence and uniqueness. It will be shown that the game may have multiple NEs, depending on the encoding order in the BD-DPC precoding design at each BS. In addition, the condition for the uniqueness of the BD-DPC multicell precoding game is generally stricter than that of the BD one. Under the IC mode, a numerical algorithm is proposed to maximize the WSR of the multicell system with BD-DPC precoding. In a conventional single-cell system, BD-DPC precoding can yield a better sum-rate performance over the BD precoding. The numerical simulations then confirm this observation for the multicell system under both IA and IC modes.

5.1 System Model

Consider a multiuser multicell downlink system with Q separate cells operating on the same frequency channel. At a particular cell, say cell-q, a multiple-antenna BS is concurrently sending independent information streams to K remote MSs, each equipped with multiple receive antennas. Let M and N be the numbers of antennas of the BS and the ith MS at cell-q, respectively. Denote $\mathbf{x}_q \in \mathbb{C}^{M \times 1}$ as the transmitted signal vector from BS-q. Assuming linear precoding at the BS, $\mathbf{x}_q$ can be represented as $\mathbf{x}_q = \sum_{i=1}^{K} \mathbf{W}_{q_i} \mathbf{s}_{q_i}$, where $\mathbf{W}_{q_i} \in \mathbb{C}^{M \times D_{q_i}}$ is the precoding matrix and $\mathbf{s}_{q_i} \in \mathbb{C}^{D_{q_i} \times 1}$ is the data symbol vector intended for MS-i. Without loss of generality, it is assumed that $\mathbb{E}\left[\mathbf{s}_{q_i} \mathbf{s}_{q_i}^H \right] = \mathbf{I}, \forall i, \forall q$.

Let $\mathbf{H}_{rq_i} \in \mathbb{C}^{N \times M_r}$ model the channel coefficients from BS-r to MS-i of cell-q, and $\mathbf{z}_{q_i}$ model the zero-mean complex additive Gaussian noise vector with an arbitrary covariance matrix $\mathbf{Z}_{q_i}$. The transmission to MS-i at cell-q can be modeled as

$$\mathbf{y}_{q_i} = \sum_{r=1}^{Q} \mathbf{H}_{rq_i} \mathbf{x}_r + \mathbf{z}_{q_i}$$

$$= \mathbf{H}_{qq_i} \mathbf{W}_{q_i} \mathbf{s}_{q_i} + \mathbf{H}_{qq_i} \sum_{j \neq i}^{K} \mathbf{W}_{q_j} \mathbf{s}_{q_j} + \sum_{r \neq q}^{Q} \mathbf{H}_{rq_i} \sum_{j=1}^{K} \mathbf{W}_{r_j} \mathbf{s}_{r_j} + \mathbf{z}_{q_i}. \quad (5.1)$$

It is observed from (5.1) that the received signal at MS-i of cell-q comprises of 4 components: the useful information signal $\mathbf{H}_{qq_i} \mathbf{W}_{q_i} \mathbf{s}_{q_i}$, the intra-cell interference $\mathbf{H}_{qq_i} \sum_{j \neq i}^{K} \mathbf{W}_{q_j} \mathbf{s}_{q_j}$, the ICI $\sum_{r \neq q}^{Q} \mathbf{H}_{rq_i} \sum_{j=1}^{K} \mathbf{W}_{r_j} \mathbf{s}_{r_j}$, and the Gaussian noise $\mathbf{z}_{q_i}$. In this work, it is assumed that each MS can measure its total interference and noise power perfectly and constantly report back to its connected BS. The BS then utilizes this information to design its precoders accordingly for its connected MSs.

In the competitive design of this system model, it is assumed that each BS only possesses full knowledge of the downlink channels to the MSs in its cell, but not the channels to the MSs in other cells. As a result, the BS cannot control its induced ICI to other cells, which is then treated as background noise at the MSs. On the contrary, in the coordinated design of this system model, the BS also possesses the CSI to the MSs in the other cells. This additional channel knowledge allows the BS to control the ICI as well. Note that the BS can always fully manage the intra-cell interference within its cell by performing certain precoding techniques on a per-cell basis. The focus of this chapter is on BD precoding which can completely suppress the intra-cell interference [20]. To implement the BD precoding on a per-cell basis, it is assumed that the total number of receive antennas at the MSs does not exceed the number of transmit antennas at their connected BS, i.e., $KN \leq M, \forall q$. If the number of receive antennas at a cell exceeds the number of transmit antennas, the BS can select a subset of MSs beforehand using low-complexity selection techniques such as [19,24].

Let $\mathbf{Q}_{q_i} = \mathbf{W}_{q_i} \mathbf{W}_{q_i}^H$ be the transmit covariance matrix intended for MS-i of cell-q, and $\mathbf{Q}_q = \{\mathbf{Q}_{q_i}\}_{i=1}^{K}$ be the precoding profile for K MSs of cell-q. Likewise, let $\mathbf{Q}_{-q} = \{\mathbf{Q}_1, \dots, \mathbf{Q}_{q-1}, \mathbf{Q}_{q+1}, \dots, \mathbf{Q}_Q\}$ denote the precoding profile of all cells except cell-q. Denote by $\mathbf{R}_{q_i}(\mathbf{Q}_{-q})$ the covariance matrix of the total IPN at MS-i of cell-q, which is defined as

$$\mathbf{R}_{q_i}(\mathbf{Q}_{-q}) = \sum_{\substack{r=1 \\ r \neq q}}^{Q} \mathbf{H}_{rq_i} \left(\sum_{j=1}^{K} \mathbf{Q}_{r_j} \right) \mathbf{H}_{rq_i}^{H} + \mathbf{Z}_{q_i}. \tag{5.2}$$

With BD precoding applied on a per-cell basis at BS-q, the achievable data rate R_{q_i} to MS-i is then given by

$$R_{q_i}(\mathbf{Q}_q, \mathbf{Q}_{-q}) = \log \left| \mathbf{I} + \mathbf{H}_{qq_i}^{H} \mathbf{R}_{q_i}^{-1}(\mathbf{Q}_{-q}) \mathbf{H}_{qq_i} \mathbf{Q}_{q_i} \right|. \tag{5.3}$$

5.2 Coordinated Multicell Block-Diagonalization Precoding

5.2.1 Problem Formulation

This section examines the multicell BD precoding under the competition mode, i.e., the IA mode, where each BS selfishly designs its BD precoders without any coordination among the cells using the game-theory framework. In particular, consider a SNG where the players are the cells and the payoff functions are the sum-rates of the cells. In each cell, the BS strategically adapts its BD precoder on a per-cell basis to greedily maximize the sum-rate to its connected MSs, subject to a constraint on its transmit power.

Let $\Omega = \{1, \ldots, Q\}$ be the set of Q players. Define $R_q(\mathbf{Q}_q, \mathbf{Q}_{-q}) = \sum_{i=1}^{K} R_{q_i}(\mathbf{Q}_q, \mathbf{Q}_{-q})$ as the payoff function of player-q. Then, given a strategy profile $\mathbf{Q}_{-q}$ from other players, player-q selfishly maximizes its payoff function by solving the following optimization problem

$$\underset{\mathbf{Q}_{q_1}, \ldots, \mathbf{Q}_{q_K}}{\text{maximize}} \quad R_q(\mathbf{Q}_q, \mathbf{Q}_{-q}) \tag{5.4}$$

$$\text{subject to} \quad \sum_{i=1}^{K} \text{Tr}\{\mathbf{Q}_{q_i}\} \leq P_q$$

$$\mathbf{H}_{qq_j} \mathbf{Q}_{q_i} \mathbf{H}_{qq_j}^{H} = \mathbf{0}, \forall j \neq i$$

$$\mathbf{Q}_{q_i} \succeq \mathbf{0}, \forall i,$$

where P_q is the power budget at BS-q. To achieve the maximum sum data-rate at cell-q, it is assumed that the IPN matrix $\mathbf{R}_{q_i}(\mathbf{Q}_{-q})$ is perfectly measured at the corresponding MS-i and reported back to its connected BS. Clearly, the optimization problem (5.4) shows that the optimal strategy of player-q does depend on the strategies of others. It is to be noted that the optimization (5.4) is carried with only local information (intra-cell CSI and signaling between the MSs and its connected BS). Thus, the BD precoding game is implemented in a fully distributed manner without any signaling exchanges among the BSs.

Due to the constraints $\mathbf{H}_{qq_i}\mathbf{Q}_{q_j}\mathbf{H}_{qq_i}^H = \mathbf{0}, \forall j \neq i$, each column of the precoder matrix $\mathbf{W}_{q_i}$ must be in the null space created by $\hat{\mathbf{H}}_{q_i} = [\mathbf{H}_{qq_1}^T, \ldots, \mathbf{H}_{qq_i-1}^T, \mathbf{H}_{qq_i+1}^T, \ldots, \mathbf{H}_{qq_K}^T]^T$. Suppose that one performs the singular value decomposition of the $(K-1)N \times M$ matrix $\hat{\mathbf{H}}_{q_i}$ as

$$\hat{\mathbf{H}}_{q_i} = \mathbf{U}_{q_i} \Sigma_{q_i} \mathbf{V}_{q_i}^H = \mathbf{U}_{q_i} \begin{bmatrix} \tilde{\Sigma}_{q_i}, & \mathbf{0} \end{bmatrix} \begin{bmatrix} \tilde{\mathbf{V}}_{q_i}^H \\ \hat{\mathbf{V}}_{q_i}^H \end{bmatrix}, \tag{5.5}$$

where $\tilde{\Sigma}_{q_i}$ is diagonal, $\mathbf{U}_{q_i}$ and $\mathbf{V}_{q_i}$ are unitary matrices, and $\hat{\mathbf{V}}_{q_i}$ is formed by the last $\hat{N} \triangleq M - (K-1)N$ columns of $\mathbf{V}_{q_i}$. Then, any precoding covariance matrix $\mathbf{Q}_{q_i}$ formulated as $\hat{\mathbf{V}}_{q_i} \mathbf{D}_{q_i} \hat{\mathbf{V}}_{q_i}^H$, where $\mathbf{D}_{q_i} \succeq \mathbf{0}$ is an arbitrary $\hat{N} \times \hat{N}$ matrix, would make $\mathbf{H}_{qq_j} \mathbf{Q}_{q_i} \mathbf{H}_{qq_j}^H = \mathbf{0}, \forall j \neq i$. Thus, the set of admissible strategies for player-q can be defined as follows:

$$\mathscr{S}_q = \left\{ \mathbf{Q}_{q_i} \in \mathbb{S}^{M \times M_q} : \mathbf{Q}_{q_i} = \hat{\mathbf{V}}_{q_i} \mathbf{D}_{q_i} \hat{\mathbf{V}}_{q_i}^H, \mathbf{D}_{q_i} \succeq \mathbf{0}, \sum_{i=1}^{K} \mathrm{Tr}\{\mathbf{D}_{q_i}\} \leq P_q \right\}. \tag{5.6}$$

Mathematically, the game has the following structure

$$\mathscr{G} = \left(\Omega, \{\mathscr{S}_q\}_{q \in \Omega}, \{R_q\}_{q \in \Omega} \right). \tag{5.7}$$

A NE of game $\mathscr{G}$ is defined when

$$R_q\left(\mathbf{Q}_q^\star, \mathbf{Q}_{-q}^\star\right) \geq R_q\left(\mathbf{Q}_q, \mathbf{Q}_{-q}^\star\right), \forall \mathbf{Q}_q \in \mathscr{S}_q, \quad \forall q \in \Omega. \tag{5.8}$$

At an NE, given the precoding strategy from other cells, a BS does not have the incentive to unilaterally change its precoding strategy, i.e., it shall achieve a lower sum-rate with the same power constraint.

5.2.2 NE Characterization

This section is to study the two most fundamental questions in analyzing a SNG: the existence and uniqueness of the game's NE. The NE characterization allows us to predict a stable outcome of the noncooperative BD precoding design in game $\mathscr{G}$. The existence of a pure NE in game $\mathscr{G}$ can be deduced straightforwardly from the work in [16] for N-person quasi-concave games. First, the strategy set $\mathscr{S}_q$ for player-q defined in (5.6) is compact and convex, $\forall q$. Second, the utility function $R_q(\mathbf{Q}_q, \mathbf{Q}_{-q})$ is a continuous function in the profile of strategies $\mathscr{S}_q$ and concave in $\mathbf{Q}_{q_1}, \ldots, \mathbf{Q}_{q_K}$. Thus, Theorem 1 in [16] indicates that there always exists at least one pure NE in game $\mathscr{G}$.

In order to study the uniqueness of an NE in game $\mathcal{G}$, the best response strategy at each player is first investigated. As defined in $\mathcal{S}_q$, the best response strategy of player-q must be in the form $\mathbf{Q}_{q_i} = \hat{\mathbf{V}}_{q_i} \mathbf{D}_{q_i} \hat{\mathbf{V}}_{q_i}^H, \forall i$. Let $\mathbf{D}_q \triangleq \mathrm{blk}\{\mathbf{D}_{q_i}\}, \mathbf{D} = \{\mathbf{D}_q\}_{q\in\Omega}$. Then, the best response strategy $\mathbf{D}_q$ at BS-q can be obtained from the following optimization problem

$$\underset{\mathbf{D}_{q_1},\dots,\mathbf{D}_{q_K}}{\text{maximize}} \sum_{i=1}^{K} \log \left| \mathbf{I} + \hat{\mathbf{V}}_{q_i}^H \mathbf{H}_{qq_i}^H \hat{\mathbf{R}}_{q_i}^{-1}(\mathbf{D}_{-q})\mathbf{H}_{qq_i}\hat{\mathbf{V}}_{q_i} \mathbf{D}_{q_i} \right| \tag{5.9}$$

$$\text{subject to} \sum_{i=1}^{K} \mathrm{Tr}\{\mathbf{D}_{q_i}\} \le P_q$$

$$\mathbf{D}_{q_i} \succeq \mathbf{0}, \forall i,$$

where $\hat{\mathbf{R}}_{q_i}(\mathbf{D}_{-q})$ is defined as

$$\hat{\mathbf{R}}_{q_i}(\mathbf{D}_{-q}) = \mathbf{R}_{q_i}(\mathbf{Q}_{-q}) = \sum_{r \neq q}^{Q} \mathbf{H}_{rq_i} \left(\sum_{j=1}^{K} \hat{\mathbf{V}}_{r_j} \mathbf{D}_{r_j} \hat{\mathbf{V}}_{r_j}^H \right) \mathbf{H}_{rq_i}^H + \mathbf{Z}_{q_i}. \tag{5.10}$$

By eigen-decomposing $\hat{\mathbf{V}}_{q_i}^H \mathbf{H}_{qq_i}^H \hat{\mathbf{R}}_{q_i}^{-1}(\mathbf{D}_{-q})\mathbf{H}_{qq_i}\hat{\mathbf{V}}_{q_i} = \hat{\mathbf{U}}_{q_i} \Lambda_{q_i} \hat{\mathbf{U}}_{q_i}^H$, the optimal solution to problem (5.9) can be easily obtained from the WF procedure

$$\mathbf{D}_{q_i} \triangleq \mathbf{WF}_{q_i}(\mathbf{D}_{-q}) = \hat{\mathbf{U}}_{q_i} \left[\mu_q \mathbf{I} - \Lambda_{q_i}^{-1} \right]^+ \hat{\mathbf{U}}_{q_i}^H, \tag{5.11}$$

where the water-level μ_q is adjusted to meet the power constraint $\sum_{i=1}^{K} \mathrm{Tr} \left\{ \left[\mu_q \mathbf{I} - \Lambda_{q_i}^{-1} \right]^+ \right\} = P_q$. Note that as $\hat{\mathbf{V}}_{q_i}$ only depends on in-cell channels at cell-q, BS-q only needs to strategically adapt its precoding matrices $\mathbf{D}_{q_i}, \forall i$ as in (5.11).

While the best response strategy of each player can be obtained in a closed-form solution in (5.11), the nonlinear structure of the WF operator is problematic in analyzing the uniqueness of the game's NE. Fortunately, the WF operator can be interpreted as a projection onto a closed set [18]. As studied in [18] for the case of single-user MIMO WF, this interpretation is also applicable to the multiuser WF case considered in problem (5.9).

Theorem 5.1. *The optimization problem* (5.9) *is equivalent to the following optimization problem*

$$\underset{\mathbf{D}_{q_1},\dots,\mathbf{D}_{q_K}}{\text{minimize}} \sum_{i=1}^{K} \left\| \mathbf{D}_{q_i} + \left(\hat{\mathbf{V}}_{q_i}^H \mathbf{H}_{qq_i}^H \hat{\mathbf{R}}_{q_i}^{-1}(\mathbf{D}_{-q})\mathbf{H}_{qq_i}\hat{\mathbf{V}}_{q_i} \right)^\dagger + c_q \mathbf{P}_{\mathcal{N}(\mathbf{H}_{qq_i}\hat{\mathbf{V}}_{q_i})} \right\|_F^2 \tag{5.12}$$

$$\text{subject to} \sum_{i=1}^{K} \mathrm{Tr}\{\mathbf{D}_{q_i}\} = P_q$$

$$\mathbf{D}_{q_i} \succeq \mathbf{0},$$

where $\mathbf{P}_{\mathcal{N}(\mathbf{H}_{qq_i}\hat{\mathbf{V}}_{q_i})}$ *is the projection onto the null space of* $\mathbf{H}_{qq_i}\hat{\mathbf{V}}_{q_i}$, *and* c_q *is an arbitrarily large constant satisfying* $c_q \geq P_q + \max_{\forall i, \forall k}[\Lambda_{q_i}]_{kk}^{-1}$.

Proof. The proof for this theorem can be found in [13].

From Theorem 5.1, the WF solution in (5.11) is indeed the solution of the optimization problem (5.12). Thus, the block-diagonal WF solution $\mathbf{WF}_q(\mathbf{D}_{-q}) \triangleq \mathrm{blk}\{\mathbf{WF}_{q_i}(\mathbf{D}_{-q})\}$, can be interpreted as a projection

$$\mathbf{WF}_q(\mathbf{D}_{-q}) = \left[-\mathrm{blk}\left\{\left(\mathbf{V}_{q_i}^H \mathbf{H}_{qq_i}^H \hat{\mathbf{R}}_{q_i}^{-1}(\mathbf{D}_{-q})\mathbf{H}_{qq_i}\hat{\mathbf{V}}_{q_i}\right)^{\dagger} + c_q \mathbf{P}_{\mathcal{N}(\mathbf{H}_{qq_i}\hat{\mathbf{V}}_{q_i})}\right\}\right]_{\mathscr{D}_q},$$

where $\mathscr{D}_q \triangleq \left\{\mathbf{D}_{q_i} \in \mathbb{S}^{\hat{N} \times \hat{N}} : \sum_{i=1}^{K} \mathrm{Tr}\{\mathbf{D}_{q_i}\} = P_q, \mathbf{D}_i \succeq \mathbf{0}, \forall i\right\}$ is a closed and convex set.

Define the multicell mapping $\mathbf{WF}(\mathbf{D}) = \{\mathbf{WF}_q(\mathbf{D}_{-q})\}_{q\in\Omega}$. Let $e_{\mathbf{WF}_q} = \|\mathbf{WF}_q(\mathbf{D}_{-q}^{(1)}) - \mathbf{WF}_q(\mathbf{D}_{-q}^{(2)})\|_F$ and $e_q = \|\mathbf{D}_q^{(1)} - \mathbf{D}_q^{(2)}\|_F$, for any given $\mathbf{D}^{(1)} \neq \mathbf{D}^{(2)}$ and $\mathbf{D}_q^{(1)}, \mathbf{D}_q^{(2)} \in \mathscr{D}_q, \forall q$. Then,

$$e_{\mathbf{WF}_q} = \left\|\left[-\mathrm{blk}\left\{\left(\hat{\mathbf{V}}_{q_i}^H \mathbf{H}_{qq_i}^H \hat{\mathbf{R}}_{q_i}^{-1}(\mathbf{D}_{-q}^{(1)})\mathbf{H}_{qq_i}\hat{\mathbf{V}}_{q_i}\right)^{\dagger} + c_q \mathbf{P}_{\mathcal{N}(\mathbf{H}_{qq_i}\hat{\mathbf{V}}_{q_i})}\right\}\right]_{\mathscr{D}_q}\right.$$

$$\left. - \left[-\mathrm{blk}\left\{\left(\hat{\mathbf{V}}_{q_i}^H \mathbf{H}_{qq_i}^H \hat{\mathbf{R}}_{q_i}^{-1}(\mathbf{D}_{-q}^{(2)})\mathbf{H}_{qq_i}\hat{\mathbf{V}}_{q_i}\right)^{\dagger} + c_q \mathbf{P}_{\mathcal{N}(\mathbf{H}_{qq_i}\hat{\mathbf{V}}_{q_i})}\right\}\right]_{\mathscr{D}_q}\right\|_F$$

$$\leq \left\|\mathrm{blk}\left\{-\left(\hat{\mathbf{V}}_{q_i}^H \mathbf{H}_{qq_i}^H \hat{\mathbf{R}}_{q_i}^{-1}(\mathbf{D}_{-q}^{(1)})\mathbf{H}_{qq_i}\hat{\mathbf{V}}_{q_i}\right)^{\dagger}\right\}\right.$$

$$\left. - \mathrm{blk}\left\{-\left(\hat{\mathbf{V}}_{q_i}^H \mathbf{H}_{qq_i}^H \hat{\mathbf{R}}_{q_i}^{-1}(\mathbf{D}_{-q}^{(2)})\mathbf{H}_{qq_i}\hat{\mathbf{V}}_{q_i}\right)^{\dagger}\right\}\right\|_F \tag{5.13a}$$

$$\leq \sum_{i=1}^{K} \left\|\left(\hat{\mathbf{V}}_{q_i}^H \mathbf{H}_{qq_i}^H \hat{\mathbf{R}}_{q_i}^{-1}(\mathbf{D}_{-q}^{(1)})\mathbf{H}_{qq_i}\hat{\mathbf{V}}_{q_i}\right)^{\dagger} - \left(\hat{\mathbf{V}}_{q_i}^H \mathbf{H}_{qq_i}^H \hat{\mathbf{R}}_{q_i}^{-1}(\mathbf{D}_{-q}^{(2)})\mathbf{H}_{qq_i}\hat{\mathbf{V}}_{q_i}\right)^{\dagger}\right\|_F \tag{5.13b}$$

$$\leq \sum_{i=1}^{K} \left\|\hat{\mathbf{V}}_{q_i}^{\dagger} \mathbf{H}_{qq_i}^{\dagger} \left(\hat{\mathbf{R}}_{q_i}^{-1}(\mathbf{D}_{-q}^{(1)}) - \hat{\mathbf{R}}_{q_i}^{-1}(\mathbf{D}_{-q}^{(2)})\right) \mathbf{H}_{qq_i}^{\dagger H} \hat{\mathbf{V}}_{q_i}^{\dagger H}\right\|_F \tag{5.13c}$$

$$= \sum_{i=1}^{K} \left\|\hat{\mathbf{V}}_{q_i}^{\dagger} \mathbf{H}_{qq_i}^{\dagger} \left[\sum_{r\neq q}^{Q} \mathbf{H}_{rq_i}\left[\sum_{j=1}^{K} \hat{\mathbf{V}}_{r_j}\left(\mathbf{D}_{r_j}^{(1)} - \mathbf{D}_{r_j}^{(2)}\right)\hat{\mathbf{V}}_{r_j}^H\right]\mathbf{H}_{rq_i}^H\right]\mathbf{H}_{qq_i}^{\dagger H}\hat{\mathbf{V}}_{q_i}^{\dagger H}\right\|_F$$

$$= \sum_{i=1}^{K} \left\|\hat{\mathbf{V}}_{q_i}^{\dagger} \mathbf{H}_{qq_i}^{\dagger} \left[\sum_{r\neq q}^{Q} \mathbf{H}_{rq_i}\hat{\mathbf{V}}_r\left(\mathbf{D}_r^{(1)} - \mathbf{D}_r^{(2)}\right)\hat{\mathbf{V}}_r^H \mathbf{H}_{rq_i}^H\right]\mathbf{H}_{qq_i}^{\dagger H}\hat{\mathbf{V}}_{q_i}^{\dagger H}\right\|_F$$

$$\leq \sum_{i=1}^{K} \sum_{r \neq q}^{Q} \left\| \hat{\mathbf{V}}_{q_i}^{\dagger} \mathbf{H}_{qq_i}^{\dagger} \mathbf{H}_{rq_i} \hat{\mathbf{V}}_r \left(\mathbf{D}_r^{(1)} - \mathbf{D}_r^{(2)} \right) \hat{\mathbf{V}}_r^{H} \mathbf{H}_{rq_i}^{H} \mathbf{H}_{qq_i}^{\dagger H} \hat{\mathbf{V}}_{q_i}^{\dagger H} \right\|_F \tag{5.13d}$$

$$\leq \sum_{i=1}^{K} \sum_{r \neq q}^{Q} \varrho \left(\hat{\mathbf{V}}_r^{H} \mathbf{H}_{rq_i}^{H} \mathbf{H}_{qq_i}^{\dagger H} \hat{\mathbf{V}}_{q_i}^{\dagger H} \hat{\mathbf{V}}_{q_i}^{\dagger} \mathbf{H}_{qq_i}^{\dagger} \mathbf{H}_{rq_i} \hat{\mathbf{V}}_r \right) \left\| \mathbf{D}_r^{(1)} - \mathbf{D}_r^{(2)} \right\|_F \tag{5.13e}$$

$$= \sum_{r \neq q}^{Q} [\mathbf{S}]_{q,r} \, e_r, \tag{5.13f}$$

where $\hat{\mathbf{V}}_r \triangleq [\hat{\mathbf{V}}_{r_1}, \dots, \hat{\mathbf{V}}_{r_K}]$ and $\mathbf{S} \in \mathbb{C}^{Q \times Q}$ is defined as

$$[\mathbf{S}]_{q,r} = \begin{cases} \sum_{i=1}^{K} \varrho \left(\hat{\mathbf{V}}_r^{H} \mathbf{H}_{rq_i}^{H} \mathbf{H}_{qq_i}^{\dagger H} \hat{\mathbf{V}}_{q_i}^{\dagger H} \hat{\mathbf{V}}_{q_i}^{\dagger} \mathbf{H}_{qq_i}^{\dagger} \mathbf{H}_{rq_i} \hat{\mathbf{V}}_r \right), & \text{if } r \neq q \\ 0, & \text{if } r = q. \end{cases} \tag{5.14}$$

Note that the inequality (5.13a) holds due to the non-expansive property of the projection onto a closed and convex set [2], inequalities (5.13b) and (5.13d) hold because $\mathbf{X} = \text{diag}\{\mathbf{X}_1, \dots, \mathbf{X}_K\}$ implies that $\|\mathbf{X}\|_F \leq \sum_{i=1}^{K} \|\mathbf{X}_i\|_F$, inequality (5.13c) holds due to the reverse order law for the Moore-Penrose pseudo-inverse [18], and inequality (5.13e) holds because the Frobenius norm is consistent [8].

Define the vectors $\mathbf{e}_{\text{WF}} = [e_{\text{WF}_1}, \dots, e_{\text{WF}_Q}]^T$ and $\mathbf{e} = [e_1, \dots, e_Q]^T$. The set of inequalities (5.13f) implies that

$$\mathbf{0} \leq \mathbf{e}_{\text{WF}} \leq \mathbf{S}\mathbf{e}. \tag{5.15}$$

Given a vector $\mathbf{w} = [w_q, \dots, w_Q]^T > 0$, the mapping $\mathbf{WF}(\mathbf{D})$ is a block-contraction with respect to the norm $\| \cdot \|_{F,\text{block}}^{\mathbf{w}}$, if there exists a non-negative constant $\alpha < 1$, such that

$$\left\| \mathbf{WF}(\mathbf{D}^{(1)}) - \mathbf{WF}(\mathbf{D}^{(2)}) \right\|_{F,\text{block}}^{\mathbf{w}} \leq \alpha \left\| \mathbf{D}^{(1)} - \mathbf{D}^{(2)} \right\|_{F,\text{block}}^{\mathbf{w}}, \quad \forall \mathbf{D}^{(1)}, \mathbf{D}^{(2)}. \tag{5.16}$$

From the inequality (5.15), one has

$$\|\mathbf{e}_{\text{WF}}\|_{\infty,\text{vec}}^{\mathbf{w}} \leq \|\mathbf{S}\mathbf{e}\|_{\infty,\text{vec}}^{\mathbf{w}} \leq \|\mathbf{S}\|_{\infty,\text{mat}}^{\mathbf{w}} \|\mathbf{e}\|_{\infty,\text{vec}}^{\mathbf{w}}, \tag{5.17}$$

as the induced ∞-norm $\| \cdot \|_{\infty,\text{mat}}^{\mathbf{w}}$ is consistent [8]. Then,

$$\begin{aligned} \left\| \mathbf{WF}(\mathbf{D}^{(1)}) - \mathbf{WF}(\mathbf{D}^{(2)}) \right\|_{F,\text{block}}^{\mathbf{w}} &= \max_{q \in \Omega} \frac{\left\| \mathbf{WF}_q(\mathbf{D}^{(1)}) - \mathbf{WF}_q(\mathbf{D}^{(2)}) \right\|_2}{w_q} \\ &= \|\mathbf{e}_{\text{WF}}\|_{\infty,\text{vec}}^{\mathbf{w}} \\ &\leq \|\mathbf{S}\|_{\infty,\text{mat}}^{\mathbf{w}} \|\mathbf{e}\|_{\infty,\text{vec}}^{\mathbf{w}} \\ &= \|\mathbf{S}\|_{\infty,\text{mat}}^{\mathbf{w}} \left\| \mathbf{D}^{(1)} - \mathbf{D}^{(2)} \right\|_{F,\text{block}}^{\mathbf{w}}. \end{aligned} \tag{5.18}$$

Thus, if $\|\mathbf{S}\|_{\infty,\mathrm{mat}}^{\mathbf{w}} < 1$, the $\mathbf{WF(D)}$ mapping is a contraction, which implies the uniqueness of the NE in game $\mathscr{G}$ [2]. In addition, the condition $\|\mathbf{S}\|_{\infty,\mathrm{mat}}^{\mathbf{w}} < 1$ is also sufficient to guarantee the convergence of the NE from any starting precoding strategy $\mathbf{D}_q \in \mathscr{D}_q$. Note that if $\mathbf{S}$ is a non-negative matrix, there always exists a positive vector $\mathbf{w}$ satisfying [2]

$$(C): \quad \|\mathbf{S}\|_{\infty,\mathrm{mat}}^{\mathbf{w}} < 1 \quad \Longleftrightarrow \quad \varrho(\mathbf{S}) < 1. \tag{5.19}$$

Remark 1. Assuming the path loss fading model in this multicell system, a physical interpretation of the sufficient condition (C) is as follows. When the intra-cell BS-MS distance gets smaller relatively to the distance between the BSs, the ICI becomes less dominant. Thus, the positive off-diagonal elements of $\mathbf{S}$ also become smaller. This results in a smaller spectral radius of $\mathbf{S}$. Therefore, as the MSs are getting closer to its connected BS, the probability of meeting condition (C) is higher, which then guarantees the uniqueness of the NE.

5.3 Coordinated Multicell Block-Diagonalization Precoding

5.3.1 Problem Formulation

In Sect. 5.2, a fully decentralized approach in the multicell BD precoding design was investigated and the NE of the system was characterized. However, it is well-known that the NE need not be Pareto-efficient [7]. Via coordination among the BSs, significant network sum-rate improvement can be obtained by jointly designing all the precoders at the same time. Nonetheless, this advantage may come with the expense of message passing among the coordinated BSs as explained later in this section. We investigate the coordinated multicell BD precoding design in order to jointly maximize the network WSR through the following optimization

$$\underset{\mathbf{Q}_1,\ldots,\mathbf{Q}_Q}{\text{maximize}} \sum_{q=1}^{Q} \omega_q \sum_{i=1}^{K} \log \left| \mathbf{I} + \mathbf{H}_{qq_i}^{H} \mathbf{R}_{q_i}^{-1} (\mathbf{Q}_{-q}) \mathbf{H}_{qq_i} \mathbf{Q}_{q_i} \right| \tag{5.20}$$

$$\text{subject to} \sum_{i=1}^{K} \mathrm{Tr}\{\mathbf{Q}_{q_i}\} \le P_q, \forall q$$

$$\mathbf{Q}_{q_i} \succeq \mathbf{0}, \ \forall i, \forall q$$

$$\mathbf{H}_{qq_j} \mathbf{Q}_{q_i} \mathbf{H}_{qq_j}^{H} = \mathbf{0}, \forall j \ne i, \forall q,$$

where $\omega_q \ge 0$ denotes the non-negative weight associated with BS-q. Since the BD constraints can be removed by formulating the precoding covariance matrix $\mathbf{Q}_{q_i}$ as $\hat{\mathbf{V}}_{q_i} \mathbf{D}_{q_i} \hat{\mathbf{V}}_{q_i}^{H}$ (where $\mathbf{D}_{q_i}$ is an arbitrary $\hat{N} \times \hat{N}$ and $\hat{\mathbf{V}}_{q_i}$ given in (5.5)), the optimization problem (5.20) can be restated as

$$\underset{\mathbf{D}_1,\dots,\mathbf{D}_Q}{\text{maximize}} \ \sum_{q=1}^{Q} \omega_q \sum_{i=1}^{K} \log \left| \mathbf{I} + \hat{\mathbf{V}}_{q_i}^H \mathbf{H}_{qq_i}^H \hat{\mathbf{R}}_{q_i}^{-1}(\mathbf{D}_{-q}) \mathbf{H}_{qq_i} \hat{\mathbf{V}}_{q_i} \mathbf{D}_{q_i} \right| \qquad (5.21)$$

$$\text{subject to} \ \sum_{i=1}^{K} \text{Tr}\{\mathbf{D}_{q_i}\} \le P_q, \ \forall q$$

$$\mathbf{D}_{q_i} \succeq \mathbf{0}, \ \forall i, \forall q.$$

It is observed that problem (5.21) is nonconvex due to presence of $\mathbf{D}_{r_j}$'s in the ICI term $\hat{\mathbf{R}}_{q_i}(\mathbf{D}_{-q})$'s. Thus, it is generally difficult and computationally complex to find the globally optimal solution to problem (5.21). Instead, this section focuses on proposing a low-complexity algorithm that can obtain at least a locally optimal solution.

5.3.2 ILA Solution Approach

This section presents a solution approach to the nonconvex problem (5.21) by considering it as a difference of convex (DC) program [1]. Specifically, by iteratively isolating and approximating the nonconvex part of the objective function into linear terms, one can decompose the DC program into multiple convex optimization problems. This DC solution approach, termed as the iterative linear approximation (ILA) algorithm, has been utilized in a recent work [9] in order to maximize the multicell network sum-rate with one MS per cell.

Denote $f_q(\mathbf{D}_q, \mathbf{D}_{-q}) = \sum_{r \ne q}^{Q} \omega_r \sum_{j=1}^{K} R_{r_j}(\mathbf{D}_q, \mathbf{D}_{-q})$ as the WSR of all cells except cell-q. Note that $f_q(\mathbf{D}_q, \mathbf{D}_{-q})$ is nonconcave in $\mathbf{D}_{q_i}, i = 1, \dots, K$. At a given value of $(\bar{\mathbf{D}}_q, \bar{\mathbf{D}}_{-q})$, approximate f_q by using the Taylor expansion of f_q around $\bar{\mathbf{D}}_{q_i}, i = 1, \dots, K$, and retaining the first linear term

$$f_q(\mathbf{D}_q, \bar{\mathbf{D}}_{-q}) \approx f_q(\bar{\mathbf{D}}_q, \bar{\mathbf{D}}_{-q}) - \sum_{i=1}^{K} \text{Tr}\left\{\mathbf{A}_{q_i}\left(\mathbf{D}_{q_i} - \bar{\mathbf{D}}_{q_i}\right)\right\}, \qquad (5.22)$$

where $\mathbf{A}_{q_i}$ is the negative partial derivative of f_q with respect to the $\mathbf{D}_{q_i}$, evaluated at $\bar{\mathbf{D}}_{q_i}$

$$\begin{aligned}
\mathbf{A}_{q_i} &= \left. -\frac{\partial f_q}{\partial \mathbf{D}_{q_i}} \right|_{\mathbf{D}_{q_i} = \bar{\mathbf{D}}_{q_i}} \\
&= \left. -\sum_{r \ne q}^{Q} \omega_r \sum_{j=1}^{K} \frac{\partial R_{r_j}}{\partial \mathbf{D}_{q_i}} \right|_{\mathbf{D}_{q_i} = \bar{\mathbf{D}}_{q_i}} \\
&= \left. \sum_{r \ne q}^{Q} \omega_r \sum_{j=1}^{K} \hat{\mathbf{V}}_{q_i}^H \mathbf{H}_{qr_j}^H \left[\hat{\mathbf{R}}_{r_j}^{-1} - \left(\hat{\mathbf{R}}_{r_j} + \mathbf{H}_{rr_j} \hat{\mathbf{V}}_{r_j} \mathbf{D}_{r_j} \hat{\mathbf{V}}_{r_j}^H \mathbf{H}_{rr_j}^H \right)^{-1} \right] \mathbf{H}_{qr_j} \hat{\mathbf{V}}_{q_i} \right|_{\mathbf{D}_{q_i} = \bar{\mathbf{D}}_{q_i}} .
\end{aligned}$$

$$(5.23)$$

Using (5.22), the network WSR around $(\bar{\mathbf{D}}_q, \bar{\mathbf{D}}_{-q})$ can be approximated as $\omega_q \sum_{i=1}^{K} R_{qi} - f_q(\bar{\mathbf{D}}_q, \bar{\mathbf{D}}_{-q}) - \sum_{i=1}^{K} \text{Tr}\{\mathbf{A}_{qi}(\mathbf{D}_{qi} - \bar{\mathbf{D}}_{qi})\}$. Omitting the deterministic terms in the objective function, the nonconvex problem (5.21) can be approximated as

$$\underset{\mathbf{D}_{q1},\ldots,\mathbf{D}_{qK}}{\text{maximize}} \quad \omega_q \sum_{i=1}^{K} \log \left| \mathbf{I} + \hat{\mathbf{V}}_{qi}^{H} \mathbf{H}_{qqi}^{H} \hat{\mathbf{R}}_{qi}^{-1}(\bar{\mathbf{D}}_{-q})\mathbf{H}_{qqi}\hat{\mathbf{V}}_{qi}\mathbf{D}_{qi} \right| - \sum_{i=1}^{K} \text{Tr}\{\mathbf{A}_{qi}\mathbf{D}_{qi}\} \quad (5.24)$$

$$\text{subject to} \quad \mathbf{D}_{qi} \succeq \mathbf{0}, \ \forall i$$

$$\sum_{i=1}^{K} \text{Tr}\{\mathbf{D}_{qi}\} \leq P_q,$$

which can be solved solely at BS-q. Thus, if the Q BSs take turns to approximate the original problem (5.21), it can be solved via Q per-cell separate problems (5.24).

It can be observed that the approximated problem (5.24) is similar to the sum-rate maximization problem with BD precoding, albeit the presence of the term $\sum_{i=1}^{K} \text{Tr}\{\mathbf{A}_{qi}\mathbf{D}_{qi}\}$. Herein, this term is the penalty charged on the ICI induced by BS-q to the MSs in other cells, whereas $\mathbf{A}_{qi}$ acts as the interference price. If the ICI penalty term is not presented, the BS would only attempt to maximize the sum-rate for its connected MSs. As a result, the multicell system is in *competition* mode, as studied in Sect. 5.2. In contrast, in *coordination* mode, each BS is doing its best in limiting the ICI induced to other cells through this ICI penalty mechanism.

Since the approximated problem (5.24) is now convex, it can be readily solved by standard convex optimization techniques. Via the Lagrangian duality, the closed-form solution to the problem can be obtained as follows. The Lagrangian of problem (5.24) can be stated as

$$\mathcal{L}_q(\mathbf{D}_{qi}, \lambda_q) = \omega_q \sum_{i=1}^{K} \log \left| \mathbf{I} + \hat{\mathbf{V}}_{qi}^{H} \mathbf{H}_{qqi}^{H} \hat{\mathbf{R}}_{qi}^{-1}(\bar{\mathbf{D}}_{-q})\mathbf{H}_{qqi}\hat{\mathbf{V}}_{qi}\mathbf{D}_{qi} \right|$$

$$- \sum_{i=1}^{K} \text{Tr}\{(\mathbf{A}_{qi} + \lambda_q\mathbf{I})\mathbf{D}_{qi}\} + \lambda_q P_q, \quad (5.25)$$

where $\lambda_q \geq 0$ is the Lagrangian multiplier associated with the power constraint. The dual function is then given by

$$g_q(\lambda_q) = \max_{\mathbf{D}_{qi} \succeq \mathbf{0}} \mathcal{L}_q(\mathbf{D}_{qi}, \lambda_q). \quad (5.26)$$

For a given λ_q, the optimal solution to the Lagrangian (5.25) is presented in the following proposition.

Proposition 5.1. *Let* $\mathbf{G}_{q_i}$ *be the generalized eigen-matrix of* $\hat{\mathbf{V}}_{q_i}^H \mathbf{H}_{qq_i}^H \hat{\mathbf{R}}_{q_i}^{-1}(\bar{\mathbf{D}}_{-q})\mathbf{H}_{qq_i}$ $\hat{\mathbf{V}}_{q_i}$ *and* $(\mathbf{A}_{q_i} + \lambda_q \mathbf{I})$. *The optimal solution, which maximizes the Lagrangian (5.25), must have the structure* $\mathbf{G}_{q_i} \mathbf{P}_{q_i} \mathbf{G}_{q_i}^H, i = 1, \ldots, K$, *where* $\mathbf{P}_{q_i}$ *is a diagonal matrix with non-negative elements.*

Proof. The proof for this proposition is similar to that of Proposition 1 in [9] for the case of single-user rate maximization with a penalty term. The detailed proof of this proposition is omitted for brevity.

Given $\mathbf{G}_{q_i}$ as the generalized eigen-matrix of $\hat{\mathbf{V}}_{q_i}^H \mathbf{H}_{qq_i}^H \hat{\mathbf{R}}_{q_i}^{-1}(\bar{\mathbf{D}}_{-q})\mathbf{H}_{qq_i} \hat{\mathbf{V}}_{q_i}$ and $(\mathbf{A}_{q_i} + \lambda_q \mathbf{I})$, one has

$$\Sigma_{q_i}^{(1)} = \mathbf{G}_{q_i}^H \hat{\mathbf{V}}_{q_i}^H \mathbf{H}_{qq_i}^H \hat{\mathbf{R}}_{q_i}^{-1}(\bar{\mathbf{D}}_{-q})\mathbf{H}_{qq_i} \hat{\mathbf{V}}_{q_i} \mathbf{G}_{q_i} \qquad (5.27)$$

$$\Sigma_{q_i}^{(2)} = \mathbf{G}_{q_i}^H (\mathbf{A}_{q_i} + \lambda_q \mathbf{I})\mathbf{G}_{q_i},$$

where $\Sigma_{q_i}^{(1)}$ and $\Sigma_{q_i}^{(2)}$ are diagonal and positive semi-definite. Thus, the maximization of the Lagrangian (5.25) becomes

$$\underset{\mathbf{P}_{q_i} \geq 0}{\text{maximize}} \ \omega_q \sum_{i=1}^{K} \log \left| \mathbf{I} + \mathbf{P}_{q_i} \Sigma_{q_i}^{(1)} \right| - \sum_{i=1}^{K} \text{Tr} \left\{ \mathbf{P}_{q_i} \Sigma_{q_i}^{(2)} \right\}, \qquad (5.28)$$

whose optimal solution can be obtained by the well-known WF structure

$$[\mathbf{P}_{q_i}^\star]_{n,n} = \left[\frac{\omega_q}{\left[\Sigma_{q_i}^{(2)}\right]_{n,n}} - \frac{1}{\left[\Sigma_{q_i}^{(1)}\right]_{n,n}} \right]^+, i = 1, \ldots, K. \qquad (5.29)$$

It remains to adjust the dual variable λ_q to impose the power constraint $\sum_{i=1}^{K} \text{Tr}\{\mathbf{P}_{q_i}^\star\} \leq P_q$ for the above WF solution. One can easily verify for the case $\lambda_q = 0$ whether $\sum_{i=1}^{K} \text{Tr}\{\mathbf{P}_{q_i}^\star\} < P_q$. If it holds, it means BS-q does not transmit at its full power limit. Otherwise, $\lambda_q > 0$ can be searched by the bisection method until $\sum_{i=1}^{K} \text{Tr}\{\mathbf{P}_{q_i}^\star\} = P_q$.

5.3.3 ILA Algorithm: Convergence and Distributed Implementation

This section addresses the convergence of the ILA algorithm and its distributed implementation. In order to solve the problem of Q cells in (5.21), the ILA algorithm requires each BS-q, $q = 1, \ldots, Q$, to update the parameters $\mathbf{A}_{q_i}$'s and sequentially take turns to solve its corresponding optimization (5.24). The convergence of the ILA algorithm is given in the following theorem.

Theorem 5.2. *The optimization (5.24) carried at any given BS-q always improves the network WSR. Thus, the Gauss-Seidel (sequential) iterative update across the Q BSs is guaranteed to converge to at least a local maximum.*

Proof. Suppose that $\mathbf{D}_q = \bar{\mathbf{D}}_q = \{\bar{\mathbf{D}}_{qi}\}_{i=1}^{K}, \forall q$ is obtained from the previous iteration, and $\mathbf{D}_q^{\star} = \{\mathbf{D}_{qi}^{\star}\}_{i=1}^{K}, \forall q$ is the optimal solution obtained from the optimization problem (5.24) at BS-q. Using a technique similar to the on applied in [9,23], it can be shown that $f_q(\mathbf{D}_q, \bar{\mathbf{D}}_{-q})$ is a convex function with respect to $\mathbf{D}_q$. Thus, by the first-order condition for the convex function $f_q(\mathbf{D}_q, \bar{\mathbf{D}}_{-q})$ [3], one has

$$f_q(\mathbf{D}_q^{\star}, \bar{\mathbf{D}}_{-q}) \geq f_q(\bar{\mathbf{D}}_q, \bar{\mathbf{D}}_{-q}) - \sum_{i=1}^{K} \text{Tr}\{\mathbf{A}_{qi}(\mathbf{D}_{qi}^{\star} - \bar{\mathbf{D}}_{qi})\}. \tag{5.30}$$

After the optimization (5.24) carried at BS-q, the network WSR is updated such that

$$\sum_{q=1}^{Q} \omega_q \sum_{i=1}^{K} R_{qi}(\mathbf{D}_q^{\star}, \bar{\mathbf{D}}_{-q})$$

$$= \omega_q \sum_{i=1}^{K} R_{qi}(\mathbf{D}_q^{\star}, \bar{\mathbf{D}}_{-q}) + f_q(\mathbf{D}_q^{\star}, \bar{\mathbf{D}}_{-q})$$

$$\geq \omega_q \sum_{i=1}^{K} R_{qi}(\mathbf{D}_q^{\star}, \bar{\mathbf{D}}_{-q}) + f_q(\bar{\mathbf{D}}_q, \bar{\mathbf{D}}_{-q}) - \sum_{i=1}^{K} \text{Tr}\{\mathbf{A}_{qi}(\mathbf{D}_{qi}^{\star} - \bar{\mathbf{D}}_{qi})\}$$

$$\geq \omega_q \sum_{i=1}^{K} R_{qi}(\bar{\mathbf{D}}_q, \bar{\mathbf{D}}_{-q}) + f_q(\bar{\mathbf{D}}_q, \bar{\mathbf{D}}_{-q}) - \sum_{i=1}^{K} \text{Tr}\{\mathbf{A}_{qi}(\bar{\mathbf{D}}_{qi} - \bar{\mathbf{D}}_{qi})\}$$

$$= \sum_{q=1}^{Q} \omega_q \sum_{i=1}^{K} R_{qi}(\bar{\mathbf{D}}_q, \bar{\mathbf{D}}_{-q}), \tag{5.31}$$

where the first inequality is due to the first-order condition in (5.30) and the second inequality is due to the fact that $\mathbf{D}_q^{\star}$ is the optimal solution of problem (5.24). Clearly, the optimization carried at a given BS-q always improves the network WSR. In the ILA algorithm, each BS, say BS-q, updates the parameters $\mathbf{A}_{qi}$'s and sequentially takes turn to solve its corresponding optimization (5.24). Since the network WSR is upper-bounded, the Gauss-Seidel (sequential) updates across the Q BSs must converge monotonically to at least a local maximum. This concludes the proof for Theorem 5.2.

As presented in Sect. 5.2, the BD precoding design for a multicell system under the IA mode can be implemented in a fully decentralized manner. Interestingly, distributed implementation can also be realized for the BD precoding design under the IC mode. Via the coordination and message exchange among the BSs, the ILA can be implemented distributively as follows. Since the optimization problem (5.24) can be executed at the corresponding BS with only local information (CSI and IPN at the connected MSs), it remains to show that the pricing factors $\mathbf{A}_{q_i}$'s can also be computed in a distributed manner through a message exchange mechanism among the BSs. It is observed from Eq. (5.23) that in order to compute $\mathbf{A}_{q_i}$, BS-q has to know the channels $\mathbf{H}_{qr_j}$'s to all the MSs in the other cells. This is an important requirement for BS-q to coordinate its induced ICI. In addition, BS-q needs to acquire the factor $\mathbf{B}_{r_j} = \hat{\mathbf{R}}_{r_j}^{-1} - \left(\hat{\mathbf{R}}_{r_j} + \mathbf{H}_{rr_j} \hat{\mathbf{V}}_{r_j} \mathbf{D}_{r_j} \hat{\mathbf{V}}_{r_j}^H \mathbf{H}_{rr_j}^H \right)^{-1}$ from other cells. Thus, it is required that each MS computes its corresponding factor $\mathbf{B}_{r_j}$ using local measurements on the total IPN in $\hat{\mathbf{R}}_{r_j}$ and the total IPN plus signal in $\hat{\mathbf{R}}_{r_j} + \mathbf{H}_{rr_j} \hat{\mathbf{V}}_{r_j} \mathbf{D}_{r_j} \hat{\mathbf{V}}_{r_j}^H \mathbf{H}_{rr_j}^H$. Each MS can calculate the factor $\mathbf{B}_{r_j}$ with local information and feed back to its connected BS. These factors $\mathbf{B}_{r_j}$'s are then exchanged among the BSs to evaluate the prices $\mathbf{A}_{q_i}$'s. This message exchange mechanism among the coordinated BSs is the distinct feature of the IC mode, compared to the IA mode.

5.4 Multicell BD-DPC Precoding: Competition and Coordination

This section considers the multicell system where each BS utilizes BD-DPC to the downlink transmissions of its connected MSs. It is well-known that DPC is the capacity-achieving encoding scheme for the multiuser broadcast channel [4,21,25]. In [4], a suboptimal and simpler zero-forcing DPC (ZF-DPC) scheme was proposed for single-antenna receivers that takes advantage of both DPC and ZF precoding. In ZF-DPC, the information signals sent to the multiple users are encoded in sequence such that the receiver at any user does not see any inter-user interference due to the use of ZF and DPC at the BS. In this work, a similar technique is applied to the encoding process at each BS. Due to the consideration of multi-antenna receivers, the technique shall be referred to as the BD-DPC precoding.

At any BS, say BS-q, denote the encoding sequence to its K connected MSs as $\boldsymbol{\pi}_q = [\pi_q(1), \ldots, \pi_q(K)]^T$. The concept of BD-DPC can be briefly explained as follows:

- BS-q freely designs the precoder $\mathbf{W}_{\pi_q(1)}$ for MS-$\pi_q(1)$.
- BS-q, having the noncausal knowledge of the codeword intended for MS-$\pi_q(1)$, uses DPC such that MS-$\pi_q(2)$ does not see the codeword for MS-$\pi_q(1)$ as interference. At the same time, the precoder $\mathbf{W}_{\pi_q(2)}$ for MS-$\pi_q(2)$ is designed on the null space caused by $\mathbf{H}_{q\pi_q(1)}$ to eliminate its induced interference to MS-$\pi_q(1)$.

- Similarly, to encode the signal for user-i, BS-q can utilize the noncausal knowledge of the codewords for MSs $\pi_q(1), \ldots, \pi_q(i-1)$, and design $\mathbf{W}_{\pi_q(i)}$ on the null space caused by $\hat{\mathbf{H}}'_{\pi_q(i)} = [\mathbf{H}_{q\pi_q(1)}, \ldots, \mathbf{H}_{q\pi_q(i-1)}]$.

5.4.1 Competitive Design

Similar to game $\mathscr{G}$ defined in Sect. 5.2, we consider a new game $\mathscr{G}'$, where each BS strategically adapts its BD-DPC precoders to maximize the sum-rate to its connected MSs. Mathematically, game $\mathscr{G}'$ can be defined as

$$\mathscr{G}' = \left(\Omega, \left\{ \mathscr{S}'_q(\pi_q) \right\}_{q \in \Omega}, \{R_q\}_{q \in \Omega} \right), \tag{5.32}$$

The set of admissible strategies $\mathscr{S}'_q(\pi_q)$ is now defined as

$$\mathscr{S}'_q(\pi_q) = \Big\{ \mathbf{Q}_{\pi_q(i)} \in \mathbb{S}^{M \times M_q} : \mathbf{Q}_{\pi_q(i)} = \hat{\mathbf{V}}_{\pi_q(i)} \mathbf{D}_{\pi_q(i)} \hat{\mathbf{V}}^H_{\pi_q(i)},$$

$$\mathbf{D}_{\pi_q(i)} \succeq 0, \sum_{i=1}^{K} \mathrm{Tr} \left\{ \mathbf{D}_{\pi_q(i)} \right\} \leq P_q \Big\}, \tag{5.33}$$

where $\hat{\mathbf{V}}_{\pi_q(i)}$ is the null space created by $\hat{\mathbf{H}}'_{\pi_q(i)}$. Due to the similarity between games $\mathscr{G}$ and $\mathscr{G}'$, the characterization for game $\mathscr{G}$ presented in Sect. 5.2.2 can be directly applied to game $\mathscr{G}'$. In particular, it can be concluded that there always exists at least one NE in game $\mathscr{G}'$ and the NE is unique if

$$(C') : \varrho(\mathbf{S}') < 1, \tag{5.34}$$

where $\mathbf{S}' \in \mathbb{C}^{Q \times Q}$ is defined as

$$[\mathbf{S}']_{q,r} = \begin{cases} \sum_{i=1}^{K} \varrho \left(\hat{\mathbf{V}}^H_{\pi_r} \mathbf{H}^H_{r\pi_q(i)} \mathbf{H}^{\dagger H}_{q\pi_q(i)} \hat{\mathbf{V}}^{\dagger H}_{\pi_q(i)} \hat{\mathbf{V}}^\dagger_{\pi_q(i)} \mathbf{H}^\dagger_{q\pi_q(i)} \mathbf{H}_{r\pi_q(i)} \hat{\mathbf{V}}_{\pi_r} \right), & \text{if } r \neq q \\ 0, & \text{if } r = q, \end{cases} \tag{5.35}$$

with $\hat{\mathbf{V}}_{\pi_r} \triangleq [\hat{\mathbf{V}}_{\pi_r(1)}, \ldots, \hat{\mathbf{V}}_{\pi_r(K)}]$.

Remark 2. Due to the dependence of the admissible strategy set $\mathscr{S}'_q(\pi_q)$ on the encoding order π_q at BS-q, the characterization of game $\mathscr{G}'$ strictly depends on the encoding order at each BS. In addition, with different encoding orders at a BS, say BS-q, the optimal strategies, which maximize the sum-rate at BS-q, are also different. The condition (C') for the uniqueness of game $\mathscr{G}'$ also depends on the encoding order at each BS-q. In fact, for any permutation in $\pi_1, \ldots, \pi_Q$, at least a different NE of game $\mathscr{G}'$ can be achieved. Given $K!$ encoding order permutations at BS-q, it can be concluded that game $\mathscr{G}'$ has at least $(K!)^Q$ NE points.

Remark 3. For a particular encoding order $\pi_1, \ldots, \pi_Q$ in game $\mathscr{G}'$, game $\mathscr{G}'$ provides a higher degree of freedom in designing the precoder at each BS. In fact, the size of matrix $\hat{\mathbf{V}}_{\pi_q(i)}$ in game $\mathscr{G}'$ is at least equal or larger than its counterpart $\hat{\mathbf{V}}_{q_i}$ in game $\mathscr{G}$. Intuitively, the off-diagonal elements of matrix $\mathbf{S}'$ are also larger than that of matrix $\mathbf{S}$. As a result, it is expected that the condition for the uniqueness of the NE in game $\mathscr{G}'$ is stricter than that in game $\mathscr{G}$.

5.4.2 Coordinated Design

This section is to investigate the implementation of BD-DPC precoding in a multicell system under the IC mode. In this case, consider the joint BD-DPC precoding design to maximize the network WSR as follows:

$$\underset{\mathbf{Q}_1, \ldots, \mathbf{Q}_Q}{\text{maximize}} \sum_{q=1}^{Q} \omega_q R_q(\mathbf{Q}_q, \mathbf{Q}_{-q}) \tag{5.36}$$

$$\text{subject to } \mathbf{Q}_q \in \mathscr{S}'_q, \forall q.$$

Similar to the optimization problem (5.20) considered in Sect. 5.3, the above problem is also nonconvex. Thus, the same ILA algorithm proposed in Sect. 5.3 can be applied to solve problem (5.36). In particular, due to monotonic convergence of the ILA algorithm, at least a locally optimal solution to the problem can be obtained. The only difference here is that the solution $\mathbf{Q}_{q_i}$ of problem (5.36) must be in the form $\hat{\mathbf{V}}_{\pi_q(i)} \mathbf{D}_{q_i} \hat{\mathbf{V}}_{\pi_q(i)}^H$, where $\hat{\mathbf{V}}_{\pi_q(i)}$ is the null space created by $\hat{\mathbf{H}}'_{\pi_q(i)}$, and $\mathbf{D}_{q_i}$ is obtained from the ILA algorithm.

5.5 Simulation Results

This section presents simulation results validating the studies on the uniqueness of an NE and the convergence to the NE in games $\mathscr{G}$ and $\mathscr{G}'$. Also presented are the sum-rates of the multicell system under the IA and IC modes when BD, BD-DPC, or DPC precoding is applied on a per-cell basis. Under the IA mode, the sum-rates obtained from the games $\mathscr{G}$ and $\mathscr{G}'$ are compared to the one obtained from the game where DPC precoding is applied at each BS. In the same system setting, the DPC precoding [4, 21] is performed on a per-cell basis in a noncooperative manner (each BS selfishly maximizes its own sum-rate) until the multicell system converges to a stable state. Similarly, under the IC mode, the network sum-rates (with equal

Fig. 5.1 A multicell system configuration with 3 cells, 3 users per cell

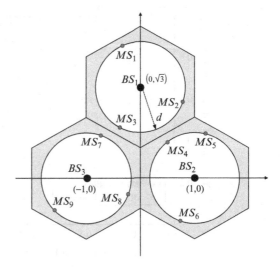

weights) obtained from the ILA algorithm for the case of BD or BD-DPC precoding design are compared to that of the DPC design.[1] For the BD-DPC or DPC precoding, a fixed encoding order from MS-1 to MS-K is assumed each BS.

Consider a 3-cell system with 3 MSs per cell sharing the same channel frequency, as illustrated in Fig. 5.1. The numbers of antennas at each BS and each MS are set at $M_q = 8$ and $N = 2$. The same power constraint $P_q = 1$ is set at each BS, unless stated otherwise. The AWGN at each MS is set as $\mathbf{Z}_{q_i} = \sigma^2\mathbf{I}$ with $\sigma^2 = 0.01$. The distance between any two BSs is normalized to 2. In each cell, the MSs are assumed to be randomly located on a circle from its connected BS with the radius of d. The channels from a BS to a MS are generated by using the path-loss model, where the path-loss exponent is set at 3. In each figure, each plotted point is obtained by averaging over 10,000 independent channel realizations.

Figure 5.2 displays the probability of the NE's uniqueness versus intra-cell BS-MS distance d by evaluating condition (C) for game $\mathscr{G}$ and (C') for game $\mathscr{G}'$. Corresponding to a small distance d is the low-ICI region (and high signal-to-interference-plus-noise (SINR) as a result). In contrast, at high d, each MS is more susceptible to higher level of ICI (low-SINR region). As observed from the figure, the uniqueness of the NE (in both games $\mathscr{G}$ and $\mathscr{G}'$) is guaranteed if the ICI is sufficiently small, as suggested in our analytical result in Sect. 5.2. In addition, the condition of the NE's uniqueness in game $\mathscr{G}'$ is much tighter than that in game $\mathscr{G}$, as analyzed in Sect. 5.4.1.

Figure 5.3 illustrates the network sum-rates obtained from the BD, BD-DPC, and DPC precoding versus the intra-cell BS-MS distance d under both IA and IC modes. It is observed from the figure that increasing the ICI powers results in

---

[1]To optimize the DPC in a multicell system under the IC mode, the numerical algorithm presented in Chap. 7 is utilized.

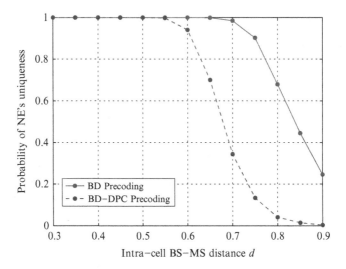

Fig. 5.2 Probability of NE's uniqueness versus the intra-cell BS-MS distance d

significant sum-rate reductions in the multicell system. Under the IA mode, while the sum-rate in the DPC precoding game is always higher than the BD and BD-DPC games due to the optimality of DPC on a per-cell basis, the performance difference is rather small. In this multicell setting, the BD and BD-DPC multicell games are much simpler to analyze than the DPC precoding game. Under the IC mode, it can be observed that the BD-DPC precoder obtains a performance very close to

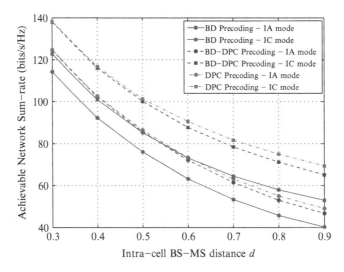

Fig. 5.3 Network sum-rates versus the intra-cell BS-MS distance d

that of the DPC precoder, especially at the high SINR (low ICI) region. Thus, it may be more advantageous to utilize BD or BD-DPC precoding over DPC because of their simpler implementation. In comparing the IC and IA modes, it can be observed that the network sum-rates can be significantly improved by coordinating the ICI. However, this performance advantage comes with the requirement of control signaling and CSI exchange among the coordinated BSs.

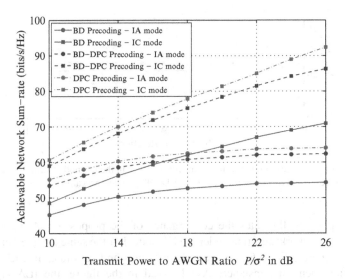

Fig. 5.4 Network sum-rates versus the transmit power to AWGN ratio at each BS for $d = 0.7$

Figure 5.4 illustrates the network sum-rates obtained from the BD, BD-DPC, and DPC precoding versus the transmit power to AWGN ratio P/σ^2 for $d = 0.7$ (assuming the same power budget P at all Q BSs). It is observed that increasing the transmit power at each BS does improve the network sum-rates in both IA and IC mode. However, at very high level of transmit power, the network sum-rates obtained from the multicell precoding games become saturated. This is due to the fact that the ICI is also increased relatively with the intra-cell information signal powers. In this case, it is desirable to coordinate and limit the amount of ICI by the IC mode. Apparently, the IC mode does perform much better than the IA mode with all 3 precoding designs at the high ICI region.

In order to illustrate the convergence of the multicell precoding games $\mathscr{G}$ and $\mathscr{G}'$, we select a specific channel realization and plot the achievable sum-rates versus the number of iterations of the two designs in Fig. 5.5. In both games, the BSs perform sequential precoder updates. The network sum-rates and the sum-rates in each cell are then plotted after each instance of updating. It is observed that both games converge very quickly in a few iterations. As expected, the BD-DPC game results in a higher network sum-rate over the BD game due to the superior performance of BD-DPC precoding over BD precoding on a per-cell basis.

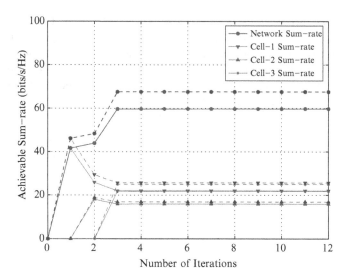

Fig. 5.5 Sum-rates versus number of iterations for $d = 0.7$ in the IA mode (*solid lines* are for the BD precoding game and *dashed lines* are for the BD-DPC precoding game)

Finally, Fig. 5.6 illustrates the convergence of the proposed ILA algorithm to maximize the network sum-rate under the IC mode. For the same channel realization utilized to generate Fig. 5.5, the network sum-rates and sum-rates in each cell are plotted after each time instance. As observed in the figure, the ILA algorithm monotonically converges with the sequential updates at the coordinated BSs for both the cases of BD and BD-DPC precoding. At each update, even though the sum-rate at one of the cells may decrease, the network sum-rate is always improved. This convergence behavior of the ILA algorithm agrees with our analysis in Theorem 5.2. As expected, the BD-DPC precoding converges to a better sum-rate than the BD precoding due to its superior performance on a per-cell basis. Compared to the convergence of the BD and BD-DPC games in Fig. 5.5, the ILA algorithm takes more iterations to converge. Nonetheless, the ILA algorithm provides better sum-rate performances for both BD and BD-DPC precoding.

5.6 Concluding Remarks

This chapter studied the multicell system with universal frequency reuse where BD or BD-DPC precoding is performed on a per-cell basis. When the multicell system is under *competition* mode, conditions on the existence and uniqueness of the multicell games' NE are investigated. Simulation results confirmed that the NE of the multicell games is unique if the ICI is sufficiently small. They also indicated that the BD-DPC multicell precoding game outperforms the BD game while achieving a

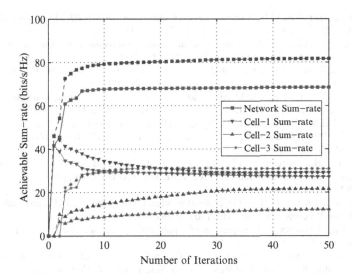

Fig. 5.6 Sum-rates versus number of iterations for $d = 0.7$ in the IC mode (*solid lines* are for the BD precoding and *dashed lines* are for the BD-DPC precoding)

sum-rate very close to that of the DPC precoding game. When the multicell system is under *coordination* mode, this chapter proposed the low-complexity and distributed ILA algorithm to obtain at least a local optimal solution to the nonconvex WSR maximization problems. Simulation results then show that the network sum-rate can be improved over the *competition* mode by coordinating the BD or BD-DPC precoders across the multicell system.

References

1. An, L.: D.C. programming for solving a class of global optimization problems via reformulation by exact penalty. In: C. Bliek, C. Jermann, A. Neumaier (eds.) Global Optimization and Constraint Satisfaction, *Lecture Notes in Computer Science*, vol. 2861, pp. 87–101. Springer Berlin / Heidelberg (2003)
2. Bertsekas, D.P., Tsitsiklis, J.N.: Parallel and Distributed Computation: Numerical Methods. Prentice-Hall, New Jersey (1989)
3. Boyd, S., Vandenberghe, L.: Convex Optimization. Cambridge University Press, United Kingdom (2004)
4. Caire, G., Shamai, S.: On the achievable throughput of a multiantenna Gaussian broadcast channel. IEEE Trans. Inform. Theory **49**(7), 1691–1706 (2003)
5. Choi, L.U., Murch, R.: A transmit preprocessing technique for multiuser MIMO systems using a decomposition approach. IEEE Trans. Wireless Commun. **3**(1), 20–24 (2004)
6. Costa, M.: Writing on dirty paper. IEEE Trans. Inform. Theory **29**(3), 439–441 (1983)
7. Dubey, P.: Inefficiency of Nash equilibria. Math. Oper. Res. **11**(1), 1–8 (1986)
8. Horn, R.A., Johnson, C.R.: Matrix Analysis. Cambridge University Press, New York (1985)

9. Kim, S.J., Giannakis, G.B.: Optimal resource allocation for MIMO ad hoc cognitive radio networks. IEEE Trans. Inform. Theory **57**(5), 3117–3131 (2011)
10. Larsson, E., Jorswieck, E.: Competition versus cooperation on the MISO interference channel. IEEE J. Select. Areas in Commun. **26**(7), 1059–1069 (2008)
11. Marsch, P., Fettweis, G.: Coordinated Multi-point in Mobile Communications: From Theory to Practice. Cambridge University Press, New York: USA (2011)
12. Nguyen, D.H.N., Le-Ngoc, T.: Multiuser downlink beamforming in multicell wireless systems: A game theoretical approach. IEEE Trans. Signal Process. **59**(7), 3326–3338 (2011)
13. Nguyen, D.H.N., Nguyen-Le, H., Le-Ngoc, T.: Block-diagonalization precoding in a multiuser multicell MIMO system: Competition and coordination. IEEE Trans. Wireless Commun. **13**(2), 968–981 (2014)
14. Nguyen-Le, H., Nguyen, D.H.N., Le-Ngoc, T.: Game-based zero-forcing precoding for multicell multiuser transmissions. In: Proc. IEEE Veh. Technol. Conf., pp. 1–5. San Francisco, CA, USA (2011)
15. Pan, Z., Wong, K.K., Ng, T.S.: Generalized multiuser orthogonal space-division multiplexing. IEEE Trans. Wireless Commun. **3**(6), 1969–1973 (2004)
16. Rosen, J.B.: Existence and uniqueness of equilibrium points for concave N-person games. Econometrica **33**(3), 520–ï£¡534 (1965)
17. Scutari, G., Palomar, D.P., Barbarossa, S.: Competitive design of multiuser MIMO system based on game theory: a unified view. IEEE J. Select. Areas in Commun. **26**(9), 1089–1102 (2008)
18. Scutari, G., Palomar, D.P., Barbarossa, S.: The MIMO iterative waterfilling algorithm. IEEE Trans. Signal Process. **57**(5), 1917–1935 (2009)
19. Shen, Z., Chen, R., Andrews, J., Heath, R., Evans, B.: Low complexity user selection algorithms for multiuser MIMO systems with block diagonalization. IEEE Trans. Signal Process. **54**(9), 3658–3663 (2006)
20. Spencer, Q., Swindlehurst, A., Haardt, M.: Zero-forcing methods for downlink spatial multiplexing in multiuser MIMO channels. IEEE Trans. Signal Process. **52**(2), 461–471 (2004)
21. Vishwanath, S., Jindal, N., Goldsmith, A.: Duality, achievable rates and sum-rate capacity of Gaussian MIMO broadcast channels. IEEE Trans. Inform. Theory **49**(10), 2658–2668 (2003)
22. Wong, K.K., Murch, R., Letaief, K.: A joint-channel diagonalization for multiuser MIMO antenna systems. IEEE Trans. Wireless Commun. **2**(4), 773–786 (2003)
23. Ye, S., Blum, R.S.: Optimized signaling for MIMO interference systems with feedback. IEEE Trans. Signal Process. **51**(11), 2839–2848 (2003)
24. Yoo, T., Goldsmith, A.: On the optimality of multiantenna broadcast scheduling using zero-forcing beamforming. IEEE J. Select. Areas in Commun. **24**(3), 528–541 (2006)
25. Yu, W., Cioffi, J.: Sum capacity of Gaussian vector broadcast channels. IEEE Trans. Inform. Theory **50**(9), 1875–1892 (2004)

Chapter 6
Sum-Rate Maximization for Multicell MIMO-MAC with IC

In Chaps. 4 and 5, it has been shown that significant power reduction or rate enhancement can be obtained by such a joint CoMP precoding design across the coordinated BSs in the downlink transmission. Similarly, in the uplink direction, it is expected that the system performance can be also improved by exploiting interference coordination among transmitting MSs. In contrast to the downlink direction, where the coordinated precoders are designed at the BSs, it is desirable for the uplink counterpart that each MS is able to determine its precoder distributively with local information only. In this case, the role of the BSs is to exchange useful control signaling to the MSs so that each MS can optimize its precoder on its own. On the other hand, the precoder at each MS has to be devised in a coordinated manner in order to maximize the link performance to its connected BS while minimizing its induced ICI to other BSs.

In this chapter, we examine a coordinated multicell system in a general setting with multiple MSs per cell, where each MS is equipped with multiple transmit antennas. In each cell, the multiple MSs concurrently transmit information signals to its connected BS, which emulates a MIMO multiple-access channel (MAC) system. Per the *coordination* mode, the BS only decodes the signals for its connected MSs by implementing the capacity-achieving decoding technique, namely successive interference cancellation (SIC). The main interest of this chapter is to jointly design the uplink precoders at the MSs with the objective of maximizing the network WSR. Since this WSR maximization problem is shown to be nonconvex, it is generally difficult and computationally complex to find its globally optimal solution.

It is known that the resource allocation problem (power allocation, precoder design, etc) for maximizing the WSR in an interference network is a challenging task. Even if there is only one MS per cell, the WSR maximization problem turns out to be nonconvex [10]. Several works in literature have examined different numerical techniques to design the transmit precoders to maximize the WSR. Specifically, the gradient projection method was applied in [10] to search for a locally optimal transmit strategy. The works in [6, 8] applied the successive convex approximation technique to decompose the original nonconvex problem into multiple convex

D.H.N. Nguyen and T. Le-Ngoc, *Wireless Coordinated Multicell Systems: Architectures and Precoding Designs*, SpringerBriefs in Computer Science, DOI 10.1007/978-3-319-06337-9_6, © The Author(s) 2014

problems, which can be solved separately at the transmitters. In particular, each transmitter optimizes its precoder to maximize its link data rate with an interference-penalty term on the interference induced to other links [6, 8]. This approach, being referred to as iterative linear approximation (ILA) [9], can be traced back to earlier works in difference of convex (DC) programming [1, 2, 5], where the nonconvex parts are linearly approximated into the penalty terms. In [6, 8, 10], by considering only one single MS per cell, the decomposed problem can be readily solved in closed-form at its corresponding BS [6].

When applying the ILA algorithm to the multicell MIMO-MAC, the approximation and decomposition step converts the nonconvex problem into a sequence of multiple MAC sum-rate maximization problems with *interference-penalty* terms. Each decomposed problem, which corresponds to the MAC in each cell, is then shown to be a convex program. Thus, it is possible to find of the optimal solution at each corresponding BS. However, due to the consideration of multiple MSs per cell, a closed-form optimal solution to the decomposed problem is not readily available. Instead, by exploring the inherently decoupled constraints for the transmit covariance matrix of each MS, we derive an equivalent optimization problem that can be solved sequentially over each variable matrix by a fast-converging algorithm. Interestingly, the decomposition in the ILA algorithm then reveals the structure of the optimal uplink precoders. In addition, the ILA solution approach also reveals the message signaling mechanism to facilitate its distributed implementation among the coordinated cells. Monotonic convergence to at least a local optimal solution is subsequently proven. Simulation results show that the proposed ILA algorithm can significantly improve the network WSR, in comparison to the multicell system with no interference coordination among the BSs.

6.1 System Model and Problem Formulation

Consider the multiuser uplink transmission of a multicell system with Q separate cells operating on the same frequency channel. In each cell, multiple MSs, each equipped with multiple transmit antennas, are sending independent data streams to its connected multiple-antenna BS. For simplicity of the presentation, it is assumed that the numbers of antennas at the BS and MS are M and N, respectively, and the number of MSs in each cell is K. Since the multicell system operates on the same frequency channel, the intended signal from a MS to a BS is now subject to the intra-cell interference from other MSs in the same cell, as well as the ICI from the MSs in other cells.

Considering the MAC at a particular cell, say cell-q, the received signal $\mathbf{y}_q$ at its BS can be modeled as

$$\mathbf{y}_q = \sum_{i=1}^{K} \mathbf{H}_{qq_i} \mathbf{x}_{q_i} + \sum_{r \neq q}^{Q} \sum_{i=1}^{K} \mathbf{H}_{qr_i} \mathbf{x}_{r_i} + \mathbf{z}_q, \qquad (6.1)$$

where $\mathbf{x}_{r_i} \in \mathbb{C}^{N \times 1}$ is the transmitted vector from N-antenna MS-i in the rth cell, $\mathbf{H}_{qr_i}$ models the channel from MS-i of cell-r to the qth BS, and $\mathbf{z}_q$ is the zero-mean additive Gaussian noise vector with the covariance matrix $\mathbf{Z}_q$.

Assuming linear precoding at each MS, the transmitted signal from MS-i in cell-q can be expressed as

$$\mathbf{x}_{q_i} = \mathbf{V}_{q_i} \mathbf{s}_{q_i}, \tag{6.2}$$

where $\mathbf{s}_{q_i} \in \mathbb{C}^{D \times 1}$ represents the information signal vector for MS-i, $\mathbf{V}_{q_i} \in \mathbb{C}^{N \times D}$ is the precoder matrix for MS-i, and $D = \min(M, N)$ is the maximum number of spatial data streams that can be supported by MS-i. It is assumed that $\mathbb{E}\left[\mathbf{s}_{q_i} \mathbf{s}_{q_i}^H\right] = \mathbf{I}$. In addition, $\mathbf{V}_{q_i}$ is constrained by transmit power limit P_{q_i}, i.e.,

$$\mathrm{Tr}\left\{\mathbf{V}_{q_i} \mathbf{V}_{q_i}^H\right\} \le P_{q_i}. \tag{6.3}$$

Let $\mathbf{X}_{q_i} = \mathbb{E}\left[\mathbf{x}_{q_i} \mathbf{x}_{q_i}^H\right] = \mathbf{V}_{q_i} \mathbf{V}_{q_i}^H$ be the transmit covariance matrix of MS-i of cell-q. Since $\mathrm{rank}\{\mathbf{X}_{q_i}\} = \mathrm{rank}\{\mathbf{V}_{q_i}\}$, should the optimization be carried over $\mathbf{X}_{q_i}$, assessing the rank of $\mathbf{X}_{q_i}$ then reveals the number of active streams supported by MS-i of cell-q. Denote $\mathbf{X}_q = \{\mathbf{X}_{q_i}\}_{i=1}^{K}$ as the uplink precoder profile of the K users in cell-q. Likewise, denote $\mathbf{X}_{-q} = (\mathbf{X}_1, \ldots, \mathbf{X}_{q-1}, \mathbf{X}_{q+1}, \ldots, \mathbf{X}_Q)$ as the precoding profile of all cells except cell-q. Denote

$$\mathbf{z}_{-q} = \sum_{r \ne q}^{Q} \sum_{i=1}^{K} \mathbf{H}_{qr_i} \mathbf{x}_{r_i} + \mathbf{z}_q \tag{6.4}$$

as the total ICI plus noise (IPN) at BS-q, and

$$\mathbf{R}_q = \mathbb{E}\left[\mathbf{z}_{-q} \mathbf{z}_{-q}^H\right] = \sum_{r \ne q}^{Q} \sum_{i=1}^{K} \mathbf{H}_{qr_i} \mathbf{X}_{r_i} \mathbf{H}_{qr_i}^H + \mathbf{Z}_q \tag{6.5}$$

as the covariance matrix of the IPN at BS-q.

In the coordinated design being considered, each BS only attempts to decode the signals from its connected MSs using the capacity-achieving multiuser decoding technique, namely successively interference cancellation (SIC) [4]. For instance, BS-q employs SIC for the transmissions from the K MSs within cell-q. Assuming the decoding order from MS-1 (first) to MS-K (last), for a certain IPN covariance $\mathbf{R}_q$, the achievable rate of user-i in cell-q can be expressed as

$$R_{q_i}(\mathbf{X}_q, \mathbf{X}_{-q}) = \log \frac{\left|\mathbf{R}_q + \sum_{j=i}^{K} \mathbf{H}_{qq_j} \mathbf{X}_{q_j} \mathbf{H}_{qq_j}^H\right|}{\left|\mathbf{R}_q + \sum_{j>i}^{K} \mathbf{H}_{qq_j} \mathbf{X}_{q_j} \mathbf{H}_{qq_j}^H\right|}, \tag{6.6}$$

where the intra-cell interference from user-1 to user-$(i-1)$ has been suppressed. Collectively, the MAC sum-rate of all K users in cell-q is given by [11]

$$R_q\left(\mathbf{X}_q, \mathbf{X}_{-q}\right) = \sum_{i=1}^{K} R_{q_i} = \log \left| \mathbf{I} + \mathbf{R}_q^{-1} \left(\sum_{i=1}^{K} \mathbf{H}_{qq_i} \mathbf{X}_{q_i} \mathbf{H}_{qq_i}^{H} \right) \right|, \quad (6.7)$$

where the denominators inside the log function are sequentially eliminated. Note that this sum-rate is obtained when BS-q does not decode the transmissions from the users in other cells. The network WSR is then given by $\sum_{q=1}^{Q} \omega_q R_q(\mathbf{X}_q, \mathbf{X}_{-q})$, where ω_q denotes the non-negative weight of cell-q. To maximize the network WSR, let us consider the following optimization

$$\underset{\mathbf{X}_1,\dots,\mathbf{X}_Q}{\text{maximize}} \sum_{q=1}^{Q} \omega_q \log \left| \mathbf{I} + \mathbf{R}_q^{-1} \left(\sum_{i=1}^{K} \mathbf{H}_{qq_i} \mathbf{X}_{q_i} \mathbf{H}_{qq_i}^{H} \right) \right| \quad (6.8)$$

$$\text{subject to } \text{Tr}\{\mathbf{X}_{q_i}\} \leq P_{q_i}, \ \forall i, \forall q$$

$$\mathbf{X}_{q_i} \succeq 0, \ \forall i, \forall q.$$

It is observed that problem (6.8) is clearly a nonconvex problem due to the presence of $\mathbf{X}_{q_i}$'s in the interference terms $\mathbf{R}_r$'s, $r \neq q$. Thus, obtaining the globally optimal solution to the problem is computationally complex and intractable for practical applications. It may also require a centralized solver unit to obtain such a solution. In this case, designing a low-complexity algorithm with distributed implementation to compute local optimizers becomes a more attractive option.

6.2   ILA Solution Approach for Multicell MIMO-MAC

This section presents a solution approach to the original nonconvex problem (6.8) by reformulating it as a DC program [1, 2, 5]. Specifically, by iteratively isolating and approximating the nonconvex part of the objective function, the DC program will be decomposed into multiple convex optimization problems, which can be solved distributively at each MS with low complexity. In addition, the iterative procedure allows the MSs to continuously refine and improve their uplink precoders, which eventually yields a local optimal solution of the original problem.

Denote $f_q(\mathbf{X}_q, \mathbf{X}_{-q}) = \sum_{r \neq q}^{Q} \omega_r R_r(\mathbf{X}_q, \mathbf{X}_{-q})$ as the WSR of all cells except cell-q so that the network WSR can be expressed as $\omega_q R_q(\mathbf{X}_q, \mathbf{X}_{-q}) + f_q(\mathbf{X}_q, \mathbf{X}_{-q})$. Since $f_q(\mathbf{X}_q, \mathbf{X}_{-q})$ is not a concave function in $\mathbf{X}_{q_i}$, we take the approximation to this term. At a given value $(\bar{\mathbf{X}}_q, \bar{\mathbf{X}}_{-q})$, after taking the Taylor expansion of f_q around $\bar{\mathbf{X}}_{q_i}$, $i = 1, \dots, K$, and retaining the first linear term, one has

$$f_q(\mathbf{X}_q, \bar{\mathbf{X}}_{-q}) \approx f_q(\bar{\mathbf{X}}_q, \bar{\mathbf{X}}_{-q}) - \sum_{i=1}^{K} \text{Tr}\left\{\mathbf{A}_{q_i}\left(\mathbf{X}_{q_i} - \bar{\mathbf{X}}_{q_i}\right)\right\}, \quad (6.9)$$

where $\mathbf{A}_{q_i}$ is the negative partial derivative of f_q with respect to $\mathbf{X}_{q_i}$, evaluated at $\mathbf{X}_{q_i} = \bar{\mathbf{X}}_{q_i}$

$$
\begin{aligned}
\mathbf{A}_{q_i} &= -\left. \frac{\partial f_q}{\partial \mathbf{X}_{q_i}} \right|_{\mathbf{X}_{q_i}=\bar{\mathbf{X}}_{q_i}} = -\sum_{r \neq q}^{Q} \omega_r \left. \frac{\partial R_r}{\partial \mathbf{X}_{q_i}} \right|_{\mathbf{X}_{q_i}=\bar{\mathbf{X}}_{q_i}} \\
&= \sum_{r \neq q}^{Q} \omega_r \mathbf{H}_{rq_i}^H \left[\mathbf{R}_r^{-1} - \left(\mathbf{R}_r + \sum_{j=1}^{K} \mathbf{H}_{rr_j} \mathbf{X}_{r_j} \mathbf{H}_{rr_j}^H \right)^{-1} \right] \mathbf{H}_{rq_i} \Bigg|_{\mathbf{X}_{q_i}=\bar{\mathbf{X}}_{q_i}} .
\end{aligned}
\tag{6.10}
$$

Using (6.9), the network WSR around $\bar{\mathbf{X}}_q$, $\omega_q R_q(\mathbf{X}_q, \bar{\mathbf{X}}_{-q}) + f_q(\mathbf{X}_q, \bar{\mathbf{X}}_{-q})$, can be approximated as $\omega_q R_q(\mathbf{X}_q, \bar{\mathbf{X}}_{-q}) - \sum_{i=1}^{K} \mathrm{Tr}\{\mathbf{A}_{q_i} \mathbf{X}_{q_i}\} + \left[f_q(\bar{\mathbf{X}}_q, \bar{\mathbf{X}}_{-q}) + \sum_{i=1}^{K} \mathrm{Tr}\{\mathbf{A}_{q_i} \bar{\mathbf{X}}_{q_i}\} \right]$. Since the term $f_q(\bar{\mathbf{X}}_q, \bar{\mathbf{X}}_{-q}) + \sum_{i=1}^{K} \mathrm{Tr}\{\mathbf{A}_{q_i} \bar{\mathbf{X}}_{q_i}\}$ is now fixed, it does not affect the maximization of the network WSR. Thus, the nonconvex problem (6.8) can be approximated as

$$
\underset{\mathbf{X}_{q_1},\dots,\mathbf{X}_{q_K}}{\text{maximize}} \; \omega_q \log \left| \mathbf{R}_q + \sum_{i=1}^{K} \mathbf{H}_{qq_i} \mathbf{X}_{q_i} \mathbf{H}_{qq_i}^H \right| - \sum_{i=1}^{K} \mathrm{Tr}\{\mathbf{A}_{q_i} \mathbf{X}_{q_i}\}
\tag{6.11}
$$

$$
\text{subject to } \mathrm{Tr}\{\mathbf{X}_{q_i}\} \leq P_{q_i}, \; \forall i
$$

$$
\mathbf{X}_{q_i} \succeq \mathbf{0}.
$$

which can be solved solely at cell-q. In other words, the optimization problem (6.8) can be approximately solved as Q per-cell separate optimization problems (6.11).

It is observed that the approximated problem (6.11) is similar to the MAC sum-rate maximization problem, studied in [11]. The difference here is the presence of the penalty term $\sum_{i=1}^{K} \mathrm{Tr}\{\mathbf{A}_{q_i} \mathbf{X}_{q_i}\}$, which encourages cell-$q$ to adopt a more cooperative precoding strategy by limiting the ICI to other cells. Should this term be absent, the multicell system is said to be in competition mode where each cell would selfishly maximize the sum-rate for its connected users only. This results in a noncooperative game among the cells, similar to the game studied in [7] for the case of one MS per cell. Some numerical results for this noncooperative design will be presented in comparison to the considered coordinated design.

Note that the decomposed problem (6.11), corresponding to the precoder design at cell-q, is now a convex program, unlike the original problem (6.8). Thus, it can be readily solved by any efficient convex optimization technique [3]. However, these direct solution approaches may require a centralized solver unit at the BS, and hence are not suitable for distributed implementation at the MSs for the MAC. Fortunately, it is observed that the constraints for each transmit covariance matrix $\mathbf{X}_{q_i}$ are inherently decoupled in problem (6.11). By exploring this decoupled structure, the optimization (6.11) can be solved sequentially over each variable matrix, like the MAC sum-rate maximization problem in [11]. More importantly, the optimization process over each variable can be performed at the corresponding MS in a fully distributed manner. These observations are elaborated in the following theorem.

Theorem 6.1. *For the K-user problem (6.11), $\{\mathbf{X}_{q_i}\}_{i=1}^{K}$ is an optimal solution if and only if $\mathbf{X}_{q_i}$ is the solution of the following optimization problem*

$$\underset{\mathbf{X}_{q_i}}{\text{maximize}}\ \omega_q \log \left| \mathbf{I} + \mathbf{R}_{q_i}^{-1}\mathbf{H}_{qq_i}\mathbf{X}_{q_i}\mathbf{H}_{qq_i}^{H} \right| - \text{Tr}\{\mathbf{A}_{q_i}\mathbf{X}_{q_i}\} \tag{6.12}$$

subject to $\text{Tr}\{\mathbf{X}_{q_i}\} \leq P_{q_i},\ \mathbf{X}_{q_i} \succeq \mathbf{0},$

where

$$\mathbf{R}_{q_i} = \mathbf{R}_q + \sum_{\substack{j=1 \\ j \neq i}}^{K}\mathbf{H}_{qq_j}\mathbf{X}_{q_j}\mathbf{H}_{qq_j}^{H} \tag{6.13}$$

is considered as noise.

Proof. The proof for this theorem follows the approach used in [11] for the sum-rate maximization in the MAC without the penalty components $\sum_{i=1}^{K}\text{Tr}\{\mathbf{A}_{q_i}\mathbf{X}_{q_i}\}$. Before proceeding to the main part of the proof, we briefly revisit the solution of problem (6.12), which was previously given in [6].

Given μ_{q_i} as the Lagrangian multiplier associated with the power constraint $\text{Tr}\{\mathbf{X}_{q_i}\} \leq P_{q_i}$, it was shown in [6] that the optimal solution $\mathbf{X}_{q_i}^{\star}$ must be in the form of

$$\mathbf{X}_{q_i}^{\star} = \mathbf{G}_{q_i}\mathbf{P}_{q_i}\mathbf{G}_{q_i}^{H}, \tag{6.14}$$

where $\mathbf{G}_{q_i}$ is the (normalized) generalized eigen-matrix of the pair of matrices of $\mathbf{H}_{qq_i}^{H}\mathbf{R}_{q_i}^{-1}\mathbf{H}_{qq_i}$ and $(\mathbf{A}_{q_i} + \mu_{q_i}\mathbf{I})$. The matrix $\mathbf{P}_{q_i}$ is a non-negative diagonal matrix, obtained from the following WF solution

$$\mathbf{P}_{q_i} = \left[\omega_q \Sigma_{q_i}^{(2)^{-1}} - \Sigma_{q_i}^{(1)^{-1}} \right]^{+}, \tag{6.15}$$

where $\Sigma_{q_i}^{(1)}$ and $\Sigma_{q_i}^{(2)}$ are diagonal matrices given by

$$\Sigma_{q_i}^{(1)} = \mathbf{G}_{q_i}^{H}\mathbf{H}_{qq_i}^{H}\mathbf{R}_{q_i}^{-1}\mathbf{H}_{qq_i}\mathbf{G}_{q_i}$$

$$\Sigma_{q_i}^{(2)} = \mathbf{G}_{q_i}^{H}(\mathbf{A}_{q_i} + \mu_{q_i}\mathbf{I})\mathbf{G}_{q_i}.$$

In this solution, the dual variable μ_{q_i}, behaving as the water-level in the WF process, is adjusted to enforce the power constraint. Utilizing this optimal solution, it can be shown that the optimal solution to the multiuser problem (6.11) is indeed a collection of solutions to individual single-user problems (6.12).

If part: Let $\{\bar{\mathbf{X}}_{q_i}\}_{i=1}^{K}$ be the optimal solution to the original K-user problem (6.11). Suppose that $\mathbf{X}_{q_i}$ is not the optimal solution of the corresponding problem (6.12) while treating $\mathbf{R}_{q_i} = \mathbf{R}_q + \sum_{j \neq i}^{K} \mathbf{H}_{qq_j} \bar{\mathbf{X}}_{q_j} \mathbf{H}_{qq_j}^{H}$ as noise. Then fixing all other covariance matrices $\bar{\mathbf{X}}_{q_j}, \forall j \neq i$, solving problem (6.12) obtains the optimal solution $\mathbf{X}_{q_i}^{\star}$. Clearly, $\mathbf{X}_{q_i}^{\star}$ strictly increases the objective function of the original problem (6.11). Thus, this contradicts with assumption on the optimality of $\{\bar{\mathbf{X}}_{q_i}\}_{i=1}^{K}$.

Only if part: Consider the partial Lagrangian of problem (6.11)

$$\mathcal{L}(\mathbf{X}_{q_i}, \boldsymbol{\mu}_q) = \sum_{i=1}^{K} \mu_{q_i} P_{q_i} + \omega_q \log \left| \mathbf{R}_q + \sum_{i=1}^{K} \mathbf{H}_{qq_i} \mathbf{X}_{q_i} \mathbf{H}_{qq_i}^{H} \right| - \sum_{i=1}^{K} \mathrm{Tr}\left\{ (\mathbf{A}_{q_i} + \mu_{q_i} \mathbf{I}) \mathbf{X}_{q_i} \right\},$$

(6.16)

where $\boldsymbol{\mu}_q = [\mu_{q_1}, \ldots, \mu_{q_K}]^{T}$ are the Lagrangian dual variables associated with the power constraints. For optimality, the solution of problem (6.11) must satisfy the following Karush–Kuhn–Tucker (KKT) conditions

$$\omega_q \mathbf{H}_{qq_i}^{H} \left(\mathbf{R}_q + \sum_{i=1}^{K} \mathbf{H}_{qq_i} \mathbf{X}_{q_i} \mathbf{H}_{qq_i}^{H} \right)^{-1} \mathbf{H}_{qq_i} = \mathbf{A}_{q_i} + \mu_{q_i} \mathbf{I}, \forall i$$

$$\mu_{q_i} \left(\mathrm{Tr}\left\{ \mathbf{X}_{q_i} \right\} - P_{q_i} \right) = 0, \forall i$$

$$\mu_{q_i} \geq 0, \forall i.$$

For the case of $K = 1$, it is straightforward to verify that the optimal solution of problem (6.12), $\mathbf{X}_{q_i}^{\star} = \mathbf{G}_{q_i} \mathbf{P}_{q_i} \mathbf{G}_{q_i}^{H}$, with $\mathbf{P}_{q_i}$ given in (6.15), satisfies the above KKT conditions. However, the KKT conditions for the single-user case are different from that of the multiuser case by the additional noise term $\sum_{j \neq i}^{K} \mathbf{H}_{qq_j} \mathbf{X}_{q_i} \mathbf{H}_{qq_j}^{H}$. Thus, if each $\mathbf{X}_{q_i}^{\star}$ satisfies the single-user condition while treating the signals of other MSs as noise, then collectively, the set of $\{\mathbf{X}_{q_i}^{\star}\}_{i=1}^{K}$ must satisfy the above KKT conditions for the multiuser case. Then, $\{\mathbf{X}_{q_i}^{\star}\}_{i=1}^{K}$ must be the optimal solution to the original problem (6.11).

Since the structure of the optimal solution to problem (6.11) has been revealed in Theorem 6.1, one can easily obtain its optimal solution by sequentially solving problem (6.12) for each user, i.e., MS-1 to MS-K in cell-q, until convergence. Note that each problem (6.12) can be effectively solved by the WF process, as presented in [6]. This sequential optimization at cell-q accounts for the *inner-loop* iterative precoder updates of the K MSs in cell-q.

For the problem of Q cells (6.8), the proposed ILA algorithm requires each cell-q, $q = 1, \ldots, Q$ to continuously update the parameters $\{\mathbf{A}_{q_i}\}_{i=1}^{K}$ and to take turns to solve its corresponding optimization (6.11). This sequential procedure accounts for the *outer-loop* iterative updates across the Q cells. The ILA algorithm for the multicell MIMO-MAC is summarized in Algorithm 6.1. The convergence of the proposed ILA algorithm is given in the following theorem.

Algorithm 6.1: ILA Algorithm for Multicell MIMO-MAC

---

1 Initialize $\{\mathbf{X}_{q_i}\}_{\forall q, \forall i}$, such that $\text{Tr}\{\mathbf{X}_{q_i}\} = P_{q_i}$;
2 **repeat**
3      $\bar{\mathbf{X}}_{q_i} \leftarrow \mathbf{X}_{q_i}$;
4   **for** $q = 1, 2, \ldots, Q$ **do**
5   Compute $\mathbf{R}_q$ with $\bar{\mathbf{X}}_{q_i}$ at BS-q and exchange among the BSs;
6   At BS-q, update the pricing matrix $\mathbf{A}_{q_i}$ at MS-i and perform;
7   **repeat**
8   **for** $i = 1, 2, \ldots, K$ **do**
9   Compute $\mathbf{R}_{q_i}$ at the BS and pass it to MS-i;
10   Perform $\underset{\mathbf{X}_{q_i}}{\text{maximize}}\ \omega_q \log\left|\mathbf{I} + \mathbf{R}_{q_i}^{-1}\mathbf{H}_{qq_i}\mathbf{X}_{q_i}\mathbf{H}_{qq_i}^H\right| - \text{Tr}\{\mathbf{A}_{q_i}\mathbf{X}_{q_i}\}$, with
                     $\text{Tr}\{\mathbf{X}_{q_i}\} \le P_{q_i}$ at MS-i;
11   **end**
12   **until** *convergence*;
13   **end**
14 **until** *convergence*;

Theorem 6.2. *The Gauss-Seidel (sequential) iterative update always improves network WSR and is guaranteed to converge to at least a local maximum.*

Proof. Similar to the approach in [6, 8], the proof for this theorem is established by showing that the network sum-rate is strictly nondecreasing after an update at any given cell. Suppose that $\mathbf{X}_q = \bar{\mathbf{X}}_q = \left\{\bar{\mathbf{X}}_{q_i}\right\}_{i=1}^K$, $\forall q$ from the previous outer-loop iteration, and $\mathbf{X}_q^\star = \left\{\mathbf{X}_{q_i}^\star\right\}_{i=1}^K$ as the optimal solution obtained at cell-q after the current outer-loop iteration.

Similar to the technique applied in [6, 10], it can be easily shown that $f_q(\mathbf{X}_q, \mathbf{X}_{-q})$ is a convex function with respect to $\mathbf{X}_{q_i} \in \mathscr{S}_{q_i} \triangleq \{\mathbf{X}_{q_i} | \mathbf{X}_{q_i} \succeq \mathbf{0}, \text{Tr}\{\mathbf{X}_{q_i}\} \le P_{q_i}\}$. Thus, by the first-order condition for the convex function f_q [3], one has

$$f_q(\mathbf{X}_q^\star, \bar{\mathbf{X}}_{-q}) \ge f_q(\bar{\mathbf{X}}_q, \bar{\mathbf{X}}_{-q}) - \sum_{i=1}^K \text{Tr}\{\mathbf{A}_{q_i}(\mathbf{X}_{q_i}^\star - \bar{\mathbf{X}}_{q_i})\} \qquad (6.17)$$

with $\mathbf{A}_{q_i}$ being defined in (6.10) at $\bar{\mathbf{X}}_{q_i}$.

After one Gauss-Seidel iteration, the network weighted sum-rate is updated such that

$$\sum_{q=1}^Q \omega_q R_q(\mathbf{X}_q^\star, \bar{\mathbf{X}}_{-q}) = \omega_q R_q(\mathbf{X}_q^\star, \bar{\mathbf{X}}_{-q}) + f_q(\mathbf{X}_q^\star, \bar{\mathbf{X}}_{-q})$$

$$\ge \omega_q R_q(\mathbf{X}_q^\star, \bar{\mathbf{X}}_{-q}) + f_q(\bar{\mathbf{X}}_q, \bar{\mathbf{X}}_{-q}) - \sum_{i=1}^K \text{Tr}\{\mathbf{A}_{q_i}(\mathbf{X}_{q_i}^\star - \bar{\mathbf{X}}_{q_i})\}$$

$$\geq \omega_q R_q(\bar{\mathbf{X}}_q, \bar{\mathbf{X}}_{-q}) + f_q(\bar{\mathbf{X}}_q, \bar{\mathbf{X}}_{-q}) - \sum_{i=1}^{K} \mathrm{Tr}\{\mathbf{A}_{qi}(\bar{\mathbf{X}}_{qi} - \bar{\mathbf{X}}_{qi})\}$$

$$= \sum_{q=1}^{Q} \omega_q R_q(\bar{\mathbf{X}}_q, \bar{\mathbf{X}}_{-q}),$$

where the first inequality is due to the one in (6.17), and the second inequality is due to the fact that $\mathbf{X}_q^*$ is the optimal solution of problem (6.11). Since the network sum-rate is upper-bounded and nondecreasing after each update, the sequential optimization (6.11) generates a Cauchy sequence that must converge to one of the local maxima.

It is worth mentioning that the proposed ILA algorithm can be executed by a central controller, which then passes the local optimal precoders to the corresponding MSs. In this case, the central controller must possess the full CSI knowledge of all channels in the network. On the other hand, it is possible to implement the proposed ILA algorithm in a distributed manner by assigning certain optimization steps in the algorithm to be performed each coordinated BS and MS.

6.3 Distributed ILA Algorithm for Multicell MIMO-MAC

In order to realize the distributed implementation of the proposed ILA algorithm, the following assumptions are required

- *Assumption 1:* Each MS, say MS-i of cell-q, knows the channel matrices $\mathbf{H}_{rqi}$'s to all the BS-r's in the network. This assumption allows the MS to control its induced ICI to other cells.
- *Assumption 2:* The coordinated BSs have reliable backhaul channels to exchange control information among themselves.
- *Assumption 3:* The channels are in block-fading or vary sufficiently slow such that they can be considered fixed during the optimization being performed.

It is to be noted that Algorithm 6.1 involves two levels of computations. At the inner-loop level, assuming $\mathbf{R}_q$ is known at BS-q and $\mathbf{A}_{qi}$ is known at MS-i, cell-q performs the corresponding optimization (6.11) autonomously using the result from Theorem 6.1. The role of BS-q is to measure the total signaling plus noise $\mathbf{R}_q + \sum_{i=1}^{K} \mathbf{H}_{qqi} \mathbf{X}_{qi} \mathbf{H}_{qqi}^H$, and then pass this value to its connected MSs. MS-i in cell-q, knowing its channel to its BS $\mathbf{H}_{qqi}$, can compute the noise component $\mathbf{R}_{qi}$. MS-i is then required to update its uplink covariance matrix by solving the optimization (6.12). This process, which corresponds to the inner-loop iterations, is performed until convergence in cell-q.

At the outer-loop level, each BS needs to exchange the data to compute the parameters $\{\mathbf{A}_{qi}\}_{i=1}^{K}$ for the next update. It is observed from Eq. (6.10) that MS-i in cell-q needs to know the channels $\mathbf{H}_{rq_i}$'s to all the BSs (per Assumption 1), as well as the pricing matrix

$$\mathbf{B}_r = \mathbf{R}_r^{-1} - \left(\mathbf{R}_r + \sum_{j=1}^{K} \mathbf{H}_{rr_j} \mathbf{X}_{r_j} \mathbf{H}_{rr_j}^{H} \right)^{-1}, \qquad (6.18)$$

in order to compute $\mathbf{A}_{q_i}$. Thus, it is required that each BS computes its corresponding price $\mathbf{B}_q, q = 1, \ldots, Q$, using local measurements on the IPN $\mathbf{R}_q$ and the total signal plus IPN $\mathbf{R}_q + \sum_{i=1}^{K} \mathbf{H}_{qq_i} \mathbf{X}_{q_i} \mathbf{H}_{qq_i}^{H}$. These factors are then exchanged among the BSs. Using the messages received from other cells, BS-q then can easily pass $\{\mathbf{B}_r\}_{r \neq q}$ to its connected MSs before the inner-loop iterative procedure. The outer-loop iteration is performed until the WSR reaches to a local maximum, as stated in Theorem 6.2.

Remark 1. It is shown in Theorem 6.2 that the proposed ILA algorithm allows the uplink precoders to be refined and improved after each Gauss-Seidel update, which ultimately converges to a local maximum. However, this update mechanism requires all the BSs to compute the pricing matrices $\mathbf{B}_q$'s and exchange them within the network after one cell updates its precoding matrices. To reduce the amount of information exchanges among the coordinated cells, the proposed algorithm can be also implemented using the Jacobi (simultaneous) iterative update. In particular, after the exchange of the pricing matrices, all the cells *simultaneously* update their precoding matrices. Although the convergence of the Jacobi update is not analytically proved, numerical simulations confirm its rapid convergence rate. Thus, in the simulations for the ILA algorithm, the Jacobi update is utilized to reduce the computational time.

Remark 2. When the multicell MIMO-MAC system operates under the *competition* mode, the update of the precoders across the Q cells also involves two levels of iterations. In an outer-loop iteration, each BS, say BS-q, needs to measure its IPN covariance matrix $\mathbf{R}_q$. In an inner-loop iteration, BS-q needs to continuously measure and pass its total signal plus IPN $\mathbf{R}_q + \sum_{i=1}^{K} \mathbf{H}_{qq_i} \mathbf{X}_{q_i} \mathbf{H}_{qq_i}^{H}$ to its K connected MSs, while the MSs at cell-q take turns to selfishly maximize the MAC sum-rate of cell-q by the IWF procedure [11]. Compared to this *competition* mode, the ILA algorithm requires the inter-BS signaling in each outer-loop iteration for exchanging the pricing matrices $\mathbf{B}_q$'s. However, the pricing matrices $\mathbf{B}_r, r \neq q$ are required to be passed from BS-q to its K connected MSs only once before the inner-loop iterations. Thus, the ILA algorithm does require a similar amount of intra-cell BS-MS signaling as the *competition* mode.

6.4 Simulation Results

This section presents simulation results on the achievable sum-rate of a multicell system in the uplink transmission under various levels of coordination and on the convergence behavior of the proposed ILA algorithm. The sum-rate performances for three operating modes are compared: (a) the *coordination* mode obtained from the proposed ILA algorithm (with equal weight for each cell), (b) the *competition* mode where each cell selfishly maximizes the sum-rate for its connected MS only, and (c) the *network MIMO* mode where the whole system is a single large MIMO MAC channel. Considered is a 3-cell system, where the distance between any two BSs is normalized to 2, as illustrated in Fig. 6.1. The number of MSs is set to 3 per cell, unless stated otherwise, and each MS is randomly located on a circle at distance d from its connected BS. The BS and MS are equipped with 4 and 2 antennas, respectively. The transmit power of each MS is limited to 1 W. The intra-cell and inter-cell channel coefficients are generated as products of two components: one accounts for the large-scale fading with a path loss exponent of 3 and one represents the small-scale fading using i.i.d. complex Gaussian random variables with zero mean and unit variance. The AWGN power spectral density σ^2 is set at 0.01 W/Hz.

We first investigate the achievable network sum-rates versus the intra-cell MS-BS distance d of the various algorithms, which are run until convergence. As d is varied, 10,000 channel realizations at each value of d are used to obtain the average network sum-rates plotted in Fig. 6.2. As shown in the figure, when the distance d becomes smaller, the network sum-rate increases in all 3 operating modes. This is due to the increase in strength of intra-cell channels and the reduction in the strength of inter-cell channels. Out of the 3 operating modes, *network MIMO* obtains the largest sum-rate, as it is the upper bound for any uplink multicell transmission scheme. It is also observed that by implementing the interference coordination among the cells using the proposed algorithm, one can improve the network sum-rate by 5–15 b/s/Hz over the *competition* mode, especially in the high ICI region (large d). Note that the performances of the coordination mode are obtained from the same initialization with $\mathbf{X}_{q_i} = (P_{q_i}/N)\mathbf{I}$.

As the proposed ILA algorithm does not guarantee a globally optimal performance, it is interesting to investigate the effects of the starting point on their achieved network sum-rate. For this, we run simulations for 10 different randomly generated starting points and record the best sum-rate result out of the 10 fully converged maxima for each algorithm. The resulting plot designated by *ILA-10_random* in Fig. 6.3 show a slightly increased network sum-rate as compared to the case with identity matrix starting point in Fig. 6.2. This close performance indicates that the identity matrix is a good and simple choice for the starting point. As previously discussed, both the proposed ILA algorithm exhibits the monotonic convergence, but require the coordinated cells to exchange signaling information after each outer-loop iteration. For practical implementation, it may be desirable to limit the number of iterations in order to reduce the amount of signaling exchange and it is interesting to understand the effect of number of

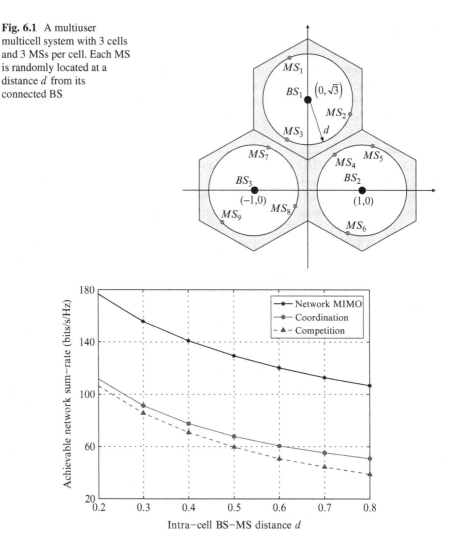

Fig. 6.1 A multiuser multicell system with 3 cells and 3 MSs per cell. Each MS is randomly located at a distance d from its connected BS

Fig. 6.2 Network sum-rates under the considered operating modes

iterations on their performance. For illustration, included in Fig. 6.3 are the plots *ILA-10_iteration* and *Competition-10_iteration*, representing the achieved network sum-rates after 10 outer-loop iterations of the ILA algorithms and competition mode, respectively. Clearly, the ILA algorithm outperforms the competition mode after just 10 iterations.

To investigate further their convergence behavior, we obtain simulation results for a specific channel realization with $d = 0.5$ and plot the network sum-rates achieved after each outer-loop iteration by ILA using Jacobi and Gauss-Seidel updates, and competition (for comparison). As observed in Fig. 6.4, the coordination

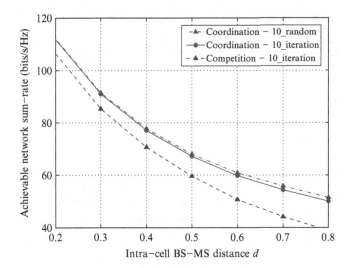

Fig. 6.3 Network sum-rates under the *coordination* mode, obtained from the ILA algorithm with 10 random starting points or with 10 outer-loop iterations

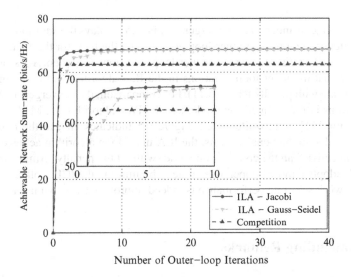

Fig. 6.4 Convergence of the proposed ILA algorithm to maximize the network sum-rate with the *coordination*

mode, obtained by the ILA algorithm, offers higher network sum-rate than the competition mode. It is also observed that the ILA algorithm does monotonically converge with both Jacobi and Gauss-Seidel updates. These convergence behavior of the ILA algorithms agree with convergence analysis in Sect. 6.3.

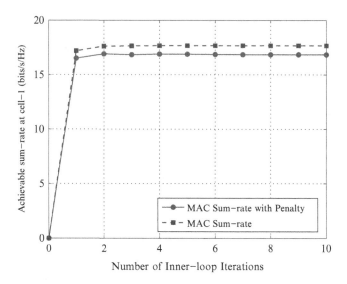

Fig. 6.5 Convergence of the proposed iterative algorithm to solve Problem (6.11)

To investigate its inner-loop convergence, Fig. 6.5 displays the sum-rate achieved at cell-1 after each number of inner-loop iterations by the ILA, for the same channel realization used in Fig. 6.4, and at outer-loop iteration #2 (when $\mathbf{A}_{q_i}$'s are non-zero). Note that the inner-loop iterations in the ILA algorithm is to maximize the MAC sum-rate with penalty terms (6.11). For comparison, the convergence of inner-loop iterations in the competition mode, i.e., the IWF algorithm for MAC sum-rate maximization [11], is also illustrated. Figure 6.5 indicates that, at the outer-loop iteration #2, due to the penalty terms, the ILA inner-loop algorithm achieves lower sum-rate in cell-1 than the competition mode (without the penalty terms). However, as the cell adopts a more cooperative strategy by limiting the ICI to other cells, the overall network sum-rate performance is indeed improved, as shown in Fig. 6.4.

6.5 Concluding Remarks

This chapter examined the problem of WSR maximization in the multicell MIMO MAC. Under the coordination mode among the multiple cells, the network WSR maximization problem was shown to be nonconvex. The ILA solution approach was then proposed to approximate and transform the original nonconvex problem into convex optimization ones, which can be solved distributed at each MS. Simulations confirmed the convergence analysis of the proposed algorithm and showed a significant enhancement in the network sum-rate as compared to competitive design.

References

1. An, L.: D.C. programming for solving a class of global optimization problems via reformulation by exact penalty. In: C. Bliek, C. Jermann, A. Neumaier (eds.) Global Optimization and Constraint Satisfaction, *Lecture Notes in Computer Science*, vol. 2861, pp. 87–101. Springer Berlin / Heidelberg (2003)
2. An, L.T.H., Tao, P.D.: The DC (difference of convex functions) programming and DCA revisited with DC models of real world nonconvex optimization problems. Annals of Operations Research **133**
3. Boyd, S., Vandenberghe, L.: Convex Optimization. Cambridge University Press, United Kingdom (2004)
4. Goldsmith, A.: Wireless Communications. Cambridge University Press, Cambridge, U.K. (2004)
5. Horst, R., Thoai, N.V.: DC programming: Overview. J. Optim. Theory Appl. **103**(1), 1–Ũ43 (1999)
6. Kim, S.J., Giannakis, G.B.: Optimal resource allocation for MIMO ad hoc cognitive radio networks. IEEE Trans. Inform. Theory **57**(5), 3117–3131 (2011)
7. Scutari, G., Palomar, D.P., Barbarossa, S.: Competitive design of multiuser MIMO system based on game theory: a unified view. IEEE J. Select. Areas in Commun. **26**(9), 1089–1102 (2008)
8. Shi, C., Berry, R.A., Honig, M.L.: Monotonic convergence of distributed interference pricing in wireless networks. In: Proc. IEEE Int. Symp. Inform. Theory, pp. 1619–1623. Seoul, Republic of Korea (2009)
9. Shi, Q., Razaviyayn, M., Luo, Z.Q., He, C.: An iteratively weighted MMSE approach to distributed sum-utility maximization for MIMO interfering broadcast channel. IEEE Trans. Signal Process. **59**(9), 4331–4340 (2011)
10. Ye, S., Blum, R.S.: Optimized signaling for MIMO interference systems with feedback. IEEE Trans. Signal Process. **51**(11), 2839–2848 (2003)
11. Yu, W., Rhee, W., Boyd, S., Cioffi, J.M.: Iterative water-filing for Gaussian multiple-access channels. IEEE Trans. Inform. Theory **50**(1), 145–152 (2004)

Chapter 7
Sum-Rate Maximization for Multicell MIMO-BC with IC

Optimizing the precoding designs in an interference network is a challenging task due to the nonconcavity of the WSR function. Different numerical methods for designing the precoders that maximize the WSR have been investigated in the literature [9,14,19]. Specifically, the gradient projection method was applied in [19] to search for a locally optimal transmit strategy. Successive convex approximation was applied in [9,14] to decompose the original nonconvex problem into a sequence of simpler convex problems, which can be solved separately at the transmitters. Note that these works only considered the network with one MS per cell. For a more general case of multiple MSs per cell, recent works in [8, 15, 16] studied the optimal linear precoding to maximize the WSR with per-BS constraints. Specifically, an iterative algorithm was proposed in [16] to solve the KKT conditions of the nonconvex WSR maximization problem. Another solution approach to the nonconvex WSR maximization problem is to transform it into a minimization of the weighted mean squared error (WMMSE) problem [3,12,15]. The WMMSE problem then can be solved by iteratively optimizing the weight matrices, the MMSE precoders, and the MMSE decoders [3]. Thus, by establishing the equivalence between the WSR maximization problem and the WMMSE minimization problem, a locally optimal solution to the former can be found from the solution of the latter.

In this chapter, the coordinated multicell system is considered in a general setting with multiple MSs per cell, where each BS or MS is equipped with multiple transmit antennas. In each cell, the BS concurrently transmits information signals to its connected MSs, which emulates a MIMO broadcast (MIMO-BC) system. The main focus of this chapter is to jointly optimize the precoding covariance matrices at the BSs in order to maximize the network-wide WSR under the IC mode. While most of the works considered linear precoding at each BS for the multicell MIMO-BC system [5, 8–11, 14–16, 19], the focus in this chapter is on nonlinear precoding design. Specifically, in the BC with multiple MSs per cell, each BS utilizes dirty-paper coding (DPC) to encode the data for the MSs within its cell. It is well-known that DPC is the capacity-achieving multiuser precoding technique for a single-cell system [2,4,17,20]. In this chapter, we extend the study

D.H.N. Nguyen and T. Le-Ngoc, *Wireless Coordinated Multicell Systems: Architectures and Precoding Designs*, SpringerBriefs in Computer Science, DOI 10.1007/978-3-319-06337-9__7, © The Author(s) 2014

of DPC onto the multicell system with interference coordination. This consideration potentially allows the multicell network to realize extra performance from the nonlinear precoding over the linear precoding. Since the maximization of WSR in a multicell MIMO-BC with DPC is a nonconvex problem, finding its globally optimal solution is computationally complex. This chapter then focuses on the iterative linear approximation (ILA) approach to numerically search for at least a locally optimal solution of the problem.

In the ILA solution approach, the sum-rate function at all cells except a particular cell under consideration is approximated into a linear interference penalty. Thus, maximizing the network WSR is equivalent to maximizing the BC sum-rate with DPC at the given cell while minimizing a penalty term on the ICI generated by its corresponding BS. Although this per-cell BC problem is yet to be convex, it is to shown that the problem is equivalent to the one in the multiple-access channel (MAC) via the so-called BC-MAC duality. Since the MAC problem is convex and thus optimally solved, the optimal solution to the BC problem is also obtained by the MAC-BC transformation [17]. Interestingly, it will be proved that the network WSR is always improved by optimizing the DPC precoders at any given BS. With the ILA algorithm, each BS is required to iteratively take turn and refine its precoders. We then prove the monotonic convergence of the ILA algorithm to at least a local maximum. In addition, a message exchange mechanism is then developed to facilitate the distributed implementation of the algorithm. Simulation results confirm the convergence analysis of the ILA algorithm, and show that the proposed algorithm significantly improves the network WSR, in comparison with linear precoding or with no IC between the BSs.

7.1   System Model and Problem Formulation

The system model considered in this chapter is the same as in Chap. 5. Consider the downlink transmission of a multicell system with Q separate cells operating on the same frequency. In each cell, a multiple-antenna BS concurrently sends independent data streams to multiple MSs, each equipped with multiple transmit antennas. For simplicity in presentation, it is assumed that the number of antennas at each BS and MS are M and N, respectively, and the number of MSs per cell is K. At a particular cell, say cell-q, the downlink transmission to MS-i can be modeled as

$$\mathbf{y}_{q_i} = \mathbf{H}_{qq_i} \sum_{j=1}^{K} \mathbf{x}_{q_j} + \sum_{r \neq q}^{Q} \mathbf{H}_{rq_i} \sum_{j=1}^{K} \mathbf{x}_{r_j} + \mathbf{z}_{q_i}, \qquad (7.1)$$

where $\mathbf{x}_{r_j} \in \mathbb{C}^{M \times 1}$ is the transmitted vector from the BS-r intended for its connected MS-j, $\mathbf{H}_{rq_i}$ models the channel from BS-r to MS-i of cell-q, and $\mathbf{z}_{q_i}$ is the zero-mean additive Gaussian noise vector with the covariance matrix $\mathbf{Z}_{q_i}$. Since the multicell system operates on the same frequency channel, the intended signal from

BS-q to its MS-i is now subject to the intra-cell interference from the signals intended for other co-located MSs in $\mathbf{H}_{qqi} \sum_{j \neq i}^{K} \mathbf{x}_{qj}$, as well as the ICI from other cells in $\sum_{r \neq q}^{Q} \mathbf{H}_{rqi} \left(\sum_{j=1}^{K} \mathbf{x}_{rj} \right)$.

Let $D = \min(M, N)$ be the number of data sub-streams for each MS. The transmit signal vector $\mathbf{x}_{qi}$ for MS-i can be expressed as

$$\mathbf{x}_{qi} = \mathbf{V}_{qi} \mathbf{s}_{qi}, \tag{7.2}$$

where $\mathbf{V}_{qi} \in \mathbb{C}^{M \times D}$ is the precoding matrix and $\mathbf{s}_{qi} \in \mathbb{C}^{D \times 1}$ represents the information signal vectors. Without loss of generality, it is assumed that $\mathbb{E}[\mathbf{s}_{qi} \mathbf{s}_{qi}^{H}] = \mathbf{I}$. Denote $\mathbf{Q}_{qi} = \mathbb{E} \left[\mathbf{x}_{qi} \mathbf{x}_{qi}^{H} \right] = \mathbf{V}_{qi} \mathbf{V}_{qi}^{H}$ as the transmit covariance matrix intended for MS-i of cell-q. Let $\mathbf{Q}_q = \{ \mathbf{Q}_{qi} \}_{i=1}^{K}$ be the downlink precoder profile of the K users at cell-q. Likewise, let $\mathbf{Q}_{-q} = (\mathbf{Q}_1, \ldots, \mathbf{Q}_{q-1}, \mathbf{Q}_{q+1}, \ldots, \mathbf{Q}_Q)$ denote the precoding profile of all cells except cell-q. Denote $\mathbf{z}_{-qi} = \sum_{r \neq q}^{Q} \mathbf{H}_{rqi} \sum_{j=1}^{K} \mathbf{x}_{rj} + \mathbf{z}_{qi}$ as the total ICI plus additive Gaussian noise at MS-i of cell-q, whose covariance $\mathbf{R}_{qi}$ is defined as

$$\mathbf{R}_{qi} = \mathbb{E} \left[\mathbf{z}_{-qi} \mathbf{z}_{-qi}^{H} \right] = \sum_{r \neq q}^{Q} \mathbf{H}_{rqi} \left(\sum_{j=1}^{K} \mathbf{Q}_{rj} \right) \mathbf{H}_{rqi}^{H} + \mathbf{Z}_{qi}. \tag{7.3}$$

As the multicell system operates in IC mode, each BS only attempts to encode and transmit information signals to the MSs within its cell. Unlike the BD precoding designs in Chap. 5, this chapter considers the capacity-achieving multiuser encoding technique, namely dirty-paper coding (DPC) [2, 4, 20], for the downlink transmissions from a BS to its connected MSs. At cell-q, assuming the encoding order from user-K to user-1, DPC is utilized such that the intended codeword for user-i does not see the intra-cell interference from user-$(i + 1)$ to user-K. As previously mentioned in Chap. 3, the achievable data rate at user-i of cell-q with DPC is given by

$$R_{qi}^{\mathrm{BC}}(\mathbf{Q}_q, \mathbf{Q}_{-q}) = \log \frac{\left| \mathbf{R}_{qi} + \mathbf{H}_{qqi} \left(\sum_{j=1}^{i} \mathbf{Q}_{qj} \right) \mathbf{H}_{qqi}^{H} \right|}{\left| \mathbf{R}_{qi} + \mathbf{H}_{qqi} \left(\sum_{j=1}^{i-1} \mathbf{Q}_{qj} \right) \mathbf{H}_{qqi}^{H} \right|}. \tag{7.4}$$

Let $R_q^{\mathrm{BC}} = \sum_{i=1}^{K} R_{qi}^{\mathrm{BC}}$ be the sum-rate at cell-q for its K connected MSs. Collectively, the network WSR is given by $\sum_{q=1}^{Q} \omega_q \sum_{i=1}^{K} R_{qi}^{\mathrm{BC}}(\mathbf{Q}_q, \mathbf{Q}_{-q})$, where ω_q denotes the non-negative weight of cell-q. Given P_q as the maximum transmit power at BS-q, the network WSR is maximized by the following optimization

$$\underset{\mathbf{Q}_1, \ldots, \mathbf{Q}_Q}{\text{maximize}} \quad \sum_{q=1}^{Q} \omega_q \sum_{i=1}^{K} R_{qi}^{\mathrm{BC}} \tag{7.5}$$

$$\text{subject to } \sum_{i=1}^{K} \text{Tr}\{\mathbf{Q}_{q_i}\} \leq P_q, \ \forall q$$

$$\mathbf{Q}_{q_i} \succeq \mathbf{0}, \ \forall i, \forall q.$$

Note that problem (7.5) is nonconvex because of the presence of $\mathbf{Q}_{q_i}$'s in the ICI terms $\mathbf{R}_{r_j}$'s with $r \neq q$, as well as the intra-cell interference term in $R_{q_j}^{\text{BC}}$ with $j < i$. To this end, the ILA algorithm, which has been applied in Chap. 6, is applied to the WSR maximization problem in the multicell MIMO-BC (7.5).

7.2 ILA Solution Approach for Multicell MIMO-BC

This section investigates the ILA solution approach to obtain at least a locally optimal solution to the nonconvex problem (7.5). Let $f_q(\mathbf{Q}_q, \mathbf{Q}_{-q}) = \sum_{r \neq q}^{Q} \omega_r R_r(\mathbf{Q}_q, \mathbf{Q}_{-q})$ denote the WSR of all cells except cell-q. As $f_q(\mathbf{Q}_q, \mathbf{Q}_{-q})$ is not concave in $\mathbf{Q}_{q_i}$, we shall take an approximation of f_q into a linear term. At a given value of $\bar{\mathbf{Q}}_{q_i}$, taking the Taylor expansion of f_q around $\bar{\mathbf{Q}}_{q_i}$ and retaining the first linear term, one has

$$f_q(\mathbf{Q}_q, \bar{\mathbf{Q}}_{-q}) \approx f_q(\bar{\mathbf{Q}}_q, \bar{\mathbf{Q}}_{-q}) - \sum_{i=1}^{K} \text{Tr}\{\mathbf{A}_q (\mathbf{Q}_{q_i} - \bar{\mathbf{Q}}_{q_i})\}, \tag{7.6}$$

where $\mathbf{A}_q$ is the negative partial derivative of f_q with respect to $\mathbf{Q}_{q_i}$, evaluated at $\mathbf{Q}_{q_i} = \bar{\mathbf{Q}}_{q_i}$, given by

$$\mathbf{A}_q = -\left. \frac{\partial f_q}{\partial \mathbf{Q}_{q_i}} \right|_{\mathbf{Q}_{q_i} = \bar{\mathbf{Q}}_{q_i}}$$

$$= -\sum_{r \neq q}^{Q} \omega_r \sum_{j=1}^{K} \left. \frac{\partial R_{r_j}}{\partial \mathbf{Q}_{q_i}} \right|_{\mathbf{Q}_{q_i} = \bar{\mathbf{Q}}_{q_i}}$$

$$= \sum_{r \neq q}^{Q} \omega_r \sum_{j=1}^{K} \mathbf{H}_{qr_j}^{H} \left[\left(\mathbf{R}_{r_j} + \sum_{k=1}^{j-1} \mathbf{H}_{rr_j} \mathbf{Q}_{r_k} \mathbf{H}_{rr_j}^{H} \right)^{-1} \right.$$

$$\left. - \left(\mathbf{R}_{r_j} + \sum_{k=1}^{j} \mathbf{H}_{rr_j} \mathbf{Q}_{r_k} \mathbf{H}_{rr_j}^{H} \right)^{-1} \right] \mathbf{H}_{qr_j} \left. \right|_{\mathbf{Q}_{q_i} = \bar{\mathbf{Q}}_{q_i}}. \tag{7.7}$$

Note that this partial derivative has the same form with respect to each $\mathbf{Q}_{q_1}, \ldots, \mathbf{Q}_{q_K}$, and is positive semi-definite, i.e., $\mathbf{A}_q \succeq \mathbf{0}$. Then, one can approximate problem (7.5) into a set of Q per-cell problems, where the optimization performed at cell-q is equivalent to

$$\underset{\mathbf{Q}_{q_1},...,\mathbf{Q}_{q_K}}{\text{maximize}} \ \omega_q \sum_{i=1}^{K} R_{q_i}^{\text{BC}} - \sum_{i=1}^{K} \text{Tr}\{\mathbf{A}_q \mathbf{Q}_{q_i}\} \tag{7.8}$$

$$\text{subject to} \ \sum_{i=1}^{K} \text{Tr}\{\mathbf{Q}_{q_i}\} \leq P_q$$

$$\mathbf{Q}_{q_i} \succeq \mathbf{0}, \forall i.$$

Some remarks regarding the optimization problem (7.8) are provided below:

Remark 1. Problem (7.8) is similar to the sum-rate maximization problem in the BC with DPC, studied in [7, 17, 20], albeit the presence of the penalty term $\sum_{i=1}^{K} \text{Tr}\{\mathbf{A}_q \mathbf{Q}_{q_i}\}$ charged on ICI generated by BS-q. The penalty term encourages the BS to design its precoders in a coordinated manner by controlling its induced ICI to other cells. Should the penalty term be omitted, the BS would only maximize the downlink capacity for its connected users. As a result, the precoding design in this multicell system is a noncooperative game between the BSs, where each BS acts as a rational and selfish player. This multicell precoding game is similar to the game studied in [13] for the case of 1 user per cell. It is to be noted that the study of the multicell precoding game with DPC for the case of multiple users per cell is beyond the scope of this work. Nonetheless, some numerical results for this noncooperative design will be presented for comparison to the coordinated design.

Remark 2. It is worth mentioning that [6, 22] studied the BC sum-rate maximization with strict constraints on the induced ICI $\sum_{i=1}^{K} \text{Tr}\{\mathbf{A}_q \mathbf{Q}_{q_i}\}$. The considered problem (7.8) is different from the studies in [6, 22], since it attempts to minimize the ICI penalty term with a sum-power constraint on the transmit covariances.

Remark 3. Although problem (7.8) is not a convex optimization problem, its resemblance to the BC's sum-rate maximization problem enables its transformation into a dual MAC maximization problem via the so-called BC-MAC duality. In a conventional multiuser MIMO system with the objective of maximizing the system sum-rate, BC-MAC duality was proved for the case of a single sum-power constraint [17, 18, 20], a set of linear power constraints [20, 21], and multiple general transmit covariance constraints [22]. In the following, it will be shown that the BC-MAC duality also holds for the multiuser MIMO system with the objective of maximizing the system sum-rate while minimizing the penalty term imposed on the transmit covariances. As a result, the nonconvex problem (7.8) can be optimally solved via the convex MAC problem by utilizing this BC-MAC duality.

For simplicity in presentation, the subscript representing the BS is dropped without loss of generality. We first consider the optimization (7.8) without the sum power constraint $\sum_{i=1}^{K} \text{Tr}\{\mathbf{Q}_i\} \leq P$, which can be stated as

$$\underset{Q_1,\dots,Q_K}{\text{maximize}} \ \ \omega \sum_{i=1}^{K} R_i^{\text{BC}} - \sum_{i=1}^{K} \text{Tr}\{\mathbf{A}\mathbf{Q}_i\} \tag{7.9}$$

$$\text{subject to} \ \ \mathbf{Q}_i \succeq \mathbf{0}, \ \forall i.$$

In the following, it is of interest to obtain the globally optimal solution $\mathbf{Q}_1^\star, \dots, \mathbf{Q}_K^\star$ to problem (7.9). Should $\mathbf{Q}_1^\star, \dots, \mathbf{Q}_K^\star$ meet the sum power constraint in (7.8), i.e., $\sum_{i=1}^{K} \text{Tr}\{\mathbf{Q}_i^\star\} \leq P$, they must be the global maximizer of problem (7.8) as well.

By changing the variables $\tilde{\mathbf{Q}}_i = \mathbf{A}^{1/2}\mathbf{Q}_i\mathbf{A}^{1/2}$ and denoting $\tilde{\mathbf{H}}_i = \mathbf{R}_i^{-1/2}\mathbf{H}_i\mathbf{A}^{-1/2}$, the data-rate to user-$i$ can be rewritten as

$$R_i^{\text{BC}} = \log \frac{\left| \mathbf{I} + \tilde{\mathbf{H}}_i \left(\sum_{j=1}^{i} \tilde{\mathbf{Q}}_i \right) \tilde{\mathbf{H}}_i^H \right|}{\left| \mathbf{I} + \tilde{\mathbf{H}}_i \left(\sum_{j=1}^{i-1} \tilde{\mathbf{Q}}_i \right) \tilde{\mathbf{H}}_i^H \right|}. \tag{7.10}$$

Thus, problem (7.9) is equivalent to

$$\underset{\tilde{Q}_1,\dots,\tilde{Q}_K}{\text{maximize}} \ \ \omega \sum_{i=1}^{K} R_i^{\text{BC}} - \sum_{i=1}^{K} \text{Tr}\{\tilde{\mathbf{Q}}_i\} \tag{7.11}$$

$$\text{subject to} \ \ \tilde{\mathbf{Q}}_i \succeq \mathbf{0}, \ \forall i.$$

In order to solve the nonconvex problem (7.11), the well-known BC-MAC duality is utilized as follows. Consider a dual MAC with K N-antenna MSs transmitting to an M-antenna BS, where the uplink channel from user-i to the BS is assumed to be $\tilde{\mathbf{H}}_i^H$ and background noise at the BS is AWGN with unit variance. It is assumed that the BS employs SIC to decode the signals from the K MSs. With the decoding order from user-1 to user-K, SIC ensures that the received signal from user-i is not interfered by the signals from user-1 to user-$(i-1)$. Denoting $\mathbf{X}_i$ as the uplink precoding covariance matrix at user-i, the achievable data-rate for user-i in the MAC is thus given by

$$R_i^{\text{MAC}} = \log \frac{\left| \mathbf{I} + \sum_{j=i}^{K} \tilde{\mathbf{H}}_j^H \mathbf{X}_j \tilde{\mathbf{H}}_j \right|}{\left| \mathbf{I} + \sum_{j>i}^{K} \tilde{\mathbf{H}}_j^H \mathbf{X}_j \tilde{\mathbf{H}}_j \right|}. \tag{7.12}$$

The key relationship between the BC and its dual MAC is presented in the following theorem.[1]

---

[1] The duality between the BC with a general linear constraint on $\sum_{i=1}^{K} \text{Tr}\{\mathbf{A}\mathbf{Q}_i\}$ and the MAC was established in [22] using a technique called SINR duality. In this work, a simple change of variables is applied to show this duality.

Theorem 7.1 ([17]). *For a given set of downlink covariance matrices $\tilde{\mathbf{Q}}_1, \ldots, \tilde{\mathbf{Q}}_K$ in the BC, it is always possible to find a set of uplink covariance matrices $\mathbf{X}_1, \ldots, \mathbf{X}_K$ such that $R_i^{\mathrm{MAC}} = R_i^{\mathrm{BC}}$ and $\sum_{i=1}^K \mathrm{Tr}\{\mathbf{X}_i\} = \sum_{i=1}^K \mathrm{Tr}\{\tilde{\mathbf{Q}}_i\}$ through the BC-MAC transformation. Vice versa, for a given set of uplink covariance matrices $\mathbf{X}_1, \ldots, \mathbf{X}_K$, it is always possible to find a set of downlink covariances $\tilde{\mathbf{Q}}_1, \ldots, \tilde{\mathbf{Q}}_K$ such that $R_i^{\mathrm{BC}} = R_i^{\mathrm{MAC}}$ and $\sum_{i=1}^K \mathrm{Tr}\{\tilde{\mathbf{Q}}_i\} = \sum_{i=1}^K \mathrm{Tr}\{\mathbf{X}_i\}$ through the MAC-BC transformation.*

From Theorem 7.1, instead of solving the nonconvex problem (7.8), one may consider the following optimization problem

$$\underset{\mathbf{X}_1, \ldots, \mathbf{X}_K}{\text{maximize}} \quad \omega \sum_{i=1}^K R_i^{\mathrm{MAC}} - \sum_{i=1}^K \mathrm{Tr}\{\mathbf{X}_i\} \qquad (7.13)$$

$$\text{subject to} \quad \mathbf{X}_i \succeq \mathbf{0}, \ \forall i,$$

which can be interpreted as a MAC sum-rate maximization with a penalty term on the transmit power at the MSs. Certainly, if the set $\mathbf{X}_1^\star, \ldots, \mathbf{X}_K^\star$ is optimal in (7.13), it is possible to find the set $\tilde{\mathbf{Q}}_1^\star, \ldots, \tilde{\mathbf{Q}}_K^\star$ that is optimal in (7.11) with the same maximum value. By contradiction, if $\tilde{\mathbf{Q}}_1^\star, \ldots, \tilde{\mathbf{Q}}_K^\star$ were not optimal, the BC-MAC transformation would ensure that $\mathbf{X}_1^\star, \ldots, \mathbf{X}_K^\star$ would not be optimal. Thus, the BC-MAC duality also holds for the considered problem with the objective of maximizing the sum-rate while minimizing the penalty term imposed on the transmit covariances. Consequently, by finding the optimal solution of problem (7.13), one also obtains the optimal solution of problem (7.11).

Note that the objective function in (7.13) can be simplified as

$$\omega \sum_{i=1}^K R_i^{\mathrm{MAC}} - \sum_{i=1}^K \mathrm{Tr}\{\mathbf{X}_i\} = \omega \log \left| \mathbf{I} + \sum_{i=1}^K \tilde{\mathbf{H}}_i^H \mathbf{X}_i \tilde{\mathbf{H}}_i \right| - \sum_{i=1}^K \mathrm{Tr}\{\mathbf{X}_i\}, \qquad (7.14)$$

which is concave in $\mathbf{X}_1, \ldots, \mathbf{X}_K$. Consequently, problem (7.13) is convex. In addition, the inherently decoupled constraints for each variable matrix $\mathbf{X}_i$ allows the sequential maximization of the objective function over each variable matrix [11]. More specifically, MS-i optimizes its covariance matrix $\mathbf{X}_i$ by performing

$$\underset{\mathbf{X}_i \succeq \mathbf{0}}{\text{maximize}} \quad \omega \log \left| \mathbf{I} + \left(\mathbf{I} + \sum_{j \neq i}^K \tilde{\mathbf{H}}_j^H \mathbf{X}_j \tilde{\mathbf{H}}_j \right)^{-1} \tilde{\mathbf{H}}_i^H \mathbf{X}_i \tilde{\mathbf{H}}_i \right| - \mathrm{Tr}\{\mathbf{X}_i\}, \qquad (7.15)$$

while treating the signal from other MSs as noise. Using the eigen-decomposition $\tilde{\mathbf{H}}_i \left(\mathbf{I} + \sum_{j \neq i}^K \tilde{\mathbf{H}}_j^H \mathbf{X}_j \tilde{\mathbf{H}}_j \right)^{-1} \tilde{\mathbf{H}}_i^H = \mathbf{U}_i \boldsymbol{\Sigma}_i \mathbf{U}_i^H$, the optimal solution can be obtained in closed-form as $\mathbf{X}_i = \mathbf{U}_i \left[\omega \mathbf{I} - \boldsymbol{\Sigma}_i^{-1} \right]^+ \mathbf{U}_i^H$. Each MS-$i$ can iteratively update its covariance matrix while keeping other covariance matrices fixed [11]. Note that this procedure always improves the objective function (7.14).

Since the objective function (7.14) is a subtraction of a log function of $\mathbf{X}_1, \ldots, \mathbf{X}_K$ to a linear function of $\mathbf{X}_1, \ldots, \mathbf{X}_K$, it must have an upper bound. As a result, the sequential optimization of (7.15) over $\mathbf{X}_1, \ldots, \mathbf{X}_K$ is guaranteed to monotonically converge to the optimal solution $\mathbf{X}_1^\star, \ldots, \mathbf{X}_K^\star$ of problem (7.13). Consequently, one can obtain the optimal solution $\tilde{\mathbf{Q}}_1^\star, \ldots, \tilde{\mathbf{Q}}_K^\star$ to problem (7.11) from $\mathbf{X}_1^\star, \ldots, \mathbf{X}_K^\star$ by the MAC-BC transformation [17]. The optimal solution of (7.9) is then given by $\mathbf{Q}_i^\star = \mathbf{A}^{-1/2} \tilde{\mathbf{Q}}_i^\star \mathbf{A}^{-1/2}$. As $\mathbf{X}_1^\star, \ldots, \mathbf{X}_K^\star$ is the globally optimal solution to the MAC problem (7.13), $\mathbf{Q}_1^\star, \ldots, \mathbf{Q}_K^\star$ must be the globally optimal solution to the BC problem (7.9). It is then straightforward to verify whether $\mathbf{Q}_1^\star, \ldots, \mathbf{Q}_K^\star$ meet the sum-power constraint $\sum_{i=1}^{K} \mathrm{Tr}\{\mathbf{Q}_i^\star\} \leq P$. If the constraint is not satisfied, one may consider the Lagrangian of original BC problem (7.8), which can be stated as

$$\mathscr{L}(\mathbf{Q}_1, \ldots, \mathbf{Q}_K, \lambda) = \omega \sum_{i=1}^{K} R_i^{\mathrm{BC}} - \sum_{i=1}^{K} \mathrm{Tr}\{(\mathbf{A} + \lambda\mathbf{I})\mathbf{Q}_i\} + \lambda P, \qquad (7.16)$$

where $\lambda \geq 0$ is the Lagrangian multiplier associated with the power constraint $\sum_{i=1}^{K} \mathrm{Tr}\{\mathbf{Q}_i\} \leq P$. The Lagrangian dual function is then given by

$$g(\lambda) = \sup_{\mathbf{Q}_1 \succeq \mathbf{0}, \ldots, \mathbf{Q}_K \succeq \mathbf{0}} \mathscr{L}(\mathbf{Q}_1, \ldots, \mathbf{Q}_K, \lambda) \qquad (7.17)$$

and the dual problem is defined as

$$\underset{\lambda}{\mathrm{minimize}} \ \ g(\lambda) \qquad (7.18)$$

$$\mathrm{subject\ to} \ \ \lambda \geq 0.$$

We first focus on the maximization of the Lagrangian dual function for a given λ, which can be stated as

$$\underset{\mathbf{Q}_1, \ldots, \mathbf{Q}_K}{\mathrm{maximize}} \ \ \omega \sum_{i=1}^{K} R_i^{\mathrm{BC}} - \sum_{i=1}^{K} \mathrm{Tr}\{(\mathbf{A} + \lambda\mathbf{I})\mathbf{Q}_i\} \qquad (7.19)$$

$$\mathrm{subject\ to} \ \ \mathbf{Q}_i \succeq \mathbf{0}, \ \forall i.$$

Clearly, problem (7.19) is similar to problem (7.9). Thus, one can obtain the globally optimal solution to problem (7.19) by adopting the approach in solving problem (7.9). The difference is in the change of variables where $\tilde{\mathbf{Q}}_i = (\mathbf{A} + \lambda\mathbf{I})^{1/2} \mathbf{Q}_i (\mathbf{A} + \lambda\mathbf{I})^{1/2}$ and $\tilde{\mathbf{H}}_i = \mathbf{R}_i^{-1/2} \mathbf{H}_i (\mathbf{A} + \lambda\mathbf{I})^{-1/2}$. Then, by solving the dual MAC problem (7.13) and performing the MAC-BC transformation one can obtain the globally optimal solution $\tilde{\mathbf{Q}}_1^\star, \ldots, \tilde{\mathbf{Q}}_K^\star$. Subsequently, the optimal solution to the Lagrangian dual problem (7.19) is given by $\mathbf{Q}_i^\star = (\mathbf{A} + \lambda\mathbf{I})^{-1/2} \tilde{\mathbf{Q}}_i^\star (\mathbf{A} + \lambda\mathbf{I})^{-1/2}$.

It remains to minimize $g(\lambda)$ subject to the constraint $\lambda \geq 0$ in (7.18). By the Lagrangian duality theory, $g(\lambda)$ is convex in λ [1]. However, $g(\lambda)$ may not be differentiable. Fortunately, it is possible to find the subgradient of $g(\lambda)$. Suppose that at λ, $\mathbf{Q}_1^\star, \ldots, \mathbf{Q}_K^\star$ is the optimal solution of (7.19). For any given $\lambda' > 0$, one has

$$
\begin{aligned}
g(\lambda') &= \max_{\{\mathbf{Q}_i\}} \omega \sum_{i=1}^{K} R_i^{\mathrm{BC}}(\{\mathbf{Q}_i\}) - \sum_{i=1}^{K} \mathrm{Tr}\{(\mathbf{A} + \lambda'\mathbf{I})\mathbf{Q}_i\} + \lambda' P \\
&\geq \omega \sum_{i=1}^{K} R_i^{\mathrm{BC}}(\{\mathbf{Q}_i^\star\}) - \sum_{i=1}^{K} \mathrm{Tr}\{(\mathbf{A} + \lambda'\mathbf{I})\mathbf{Q}_i^\star\} + \lambda' P \\
&= g(\lambda) + \left(P - \sum_{i=1}^{K} \mathrm{Tr}\{\mathbf{Q}_i^\star\}\right)(\lambda' - \lambda).
\end{aligned}
\tag{7.20}
$$

Thus, $P - \sum_{i=1}^{K} \mathrm{Tr}\{\mathbf{Q}_i^\star\}$ can be chosen as the subgradient of $g(\lambda)$. The subgradient search direction suggests to increase λ if $\sum_{i=1}^{K} \mathrm{Tr}\{\mathbf{Q}_i^\star\} \geq P$ or decrease λ otherwise. Since λ is searched in a one-dimensional space, the bisection method can be efficiently applied to find the optimal $\lambda^\star$. We summarize the proposed algorithm to solve the nonconvex problem (7.8) in Algorithm 7.1.

The optimality of the proposed algorithm is proved in the following theorem.

Theorem 7.2. *The proposed Algorithm 7.1 achieves the globally optimal solution to problem* (7.8).

Proof. Per the proposed Algorithm 7.1, if the obtained solution set $\mathbf{Q}_1^\star, \ldots, \mathbf{Q}_K^\star$ for the case $\lambda = 0$ meets the power constraint $\sum_{i=1}^{K} \mathrm{Tr}\{\mathbf{Q}_i^\star\} \leq P$, then $\mathbf{Q}_1^\star, \ldots, \mathbf{Q}_K^\star$ is the globally optimal solution to problem (7.8). This is due to the equivalence between the BC problem (7.11) and the MAC problem (7.13).

We now focus on the case $\lambda^\star > 0$. Suppose that the obtained solution $\mathbf{Q}_1^\star, \ldots, \mathbf{Q}_K^\star$ from the proposed algorithm is not globally optimal, and there is another solution set $\hat{\mathbf{Q}}_1, \ldots, \hat{\mathbf{Q}}_K$ satisfying the conditions:

(i) $\sum_{i=1}^{K} \mathrm{Tr}\{\hat{\mathbf{Q}}_i\} \leq P$
(ii) $\omega \sum_{i=1}^{K} R_i^{\mathrm{BC}}(\hat{\mathbf{Q}}) - \sum_{i=1}^{K} \mathrm{Tr}\{\mathbf{A}\hat{\mathbf{Q}}_i\} > \omega \sum_{i=1}^{K} R_i^{\mathrm{BC}}(\mathbf{Q}^\star) - \sum_{i=1}^{K} \mathrm{Tr}\{\mathbf{A}\mathbf{Q}_i^\star\}$.

Since $\mathbf{Q}_1^\star, \ldots, \mathbf{Q}_K^\star$ globally maximizes the Lagrangian as given in problem (7.19), one has

$$
\omega \sum_{i=1}^{K} R_i^{\mathrm{BC}}(\mathbf{Q}^\star) - \sum_{i=1}^{K} \mathrm{Tr}\{\mathbf{A}\mathbf{Q}_i^\star + \lambda^\star \mathbf{Q}_i^\star\} \geq \omega \sum_{i=1}^{K} R_i^{\mathrm{BC}}(\hat{\mathbf{Q}}) - \sum_{i=1}^{K} \mathrm{Tr}\{\mathbf{A}\hat{\mathbf{Q}}_i + \lambda^\star \hat{\mathbf{Q}}_i\}.
$$

Thus, condition (ii) then guarantees

Algorithm 7.1: Iterative Algorithm for the MIMO-BC Sum-rate Maximization
with a Penalty Term

---

1 **for** *a given $\lambda \geq 0$* **do**

2   Change the variables as $\tilde{\mathbf{Q}}_i = (\mathbf{A} + \lambda \mathbf{I})^{1/2} \mathbf{Q}_i (\mathbf{A} + \lambda \mathbf{I})^{1/2}$ and

        $\tilde{\mathbf{H}}_i = \mathbf{R}_i^{-1/2} \mathbf{H}_i (\mathbf{A} + \lambda \mathbf{I})^{-1/2}$;

3   Solve the equivalent uplink MAC problem

        $\underset{\mathbf{X}_1, \dots, \mathbf{X}_K}{\text{maximize}} \; \omega \log \left| \mathbf{I} + \sum_{i=1}^{K} \tilde{\mathbf{H}}_i^H \mathbf{X}_i \tilde{\mathbf{H}}_i \right| - \sum_{i=1}^{K} \text{Tr}\{\mathbf{X}_i\}$ by;

4   **repeat**

5   **for** $i = 1, 2, \dots, K$ **do**

6   Perform the eigen-decomposition

              $\tilde{\mathbf{H}}_i \left(\mathbf{I} + \sum_{j \neq i}^{K} \tilde{\mathbf{H}}_j^H \mathbf{X}_j \tilde{\mathbf{H}}_j \right)^{-1} \tilde{\mathbf{H}}_i^H = \mathbf{U}_i \boldsymbol{\Sigma}_i \mathbf{U}_i^H$;

7   Update $\mathbf{X}_i = \mathbf{U}_i \left[\omega \mathbf{I} - \boldsymbol{\Sigma}_i^{-1} \right]^+ \mathbf{U}_i$;

8   **end**

9   **until** *convergence to* $\mathbf{X}_1^\star, \dots, \mathbf{X}_K^\star$;

10   Compute $\mathbf{Q}_1^\star, \dots, \mathbf{Q}_K^\star$ from $\mathbf{X}_1^\star, \dots, \mathbf{X}_K^\star$ by the MAC-BC transformation;

11   Compute $\mathbf{Q}_i^\star = (\mathbf{A} + \lambda \mathbf{I})^{-1/2} \tilde{\mathbf{Q}}_i^\star (\mathbf{A} + \lambda \mathbf{I})^{-1/2}, i = 1, \dots, K$;

12 **end**

13 **case** $\lambda = 0$

14   Follow step 1 to 12 to obtain $\mathbf{Q}_i^\star$;

15   **if** $\sum_{i=1}^{K} \text{Tr}\{\mathbf{Q}_i^\star\} \leq P$ **then** stop the algorithm;

16 **case** $\lambda > 0$

17   Set $\lambda_{\min} = 0$ and $\lambda_{\max}$ large;

18   **repeat**

19        $\lambda = (\lambda_{\min} + \lambda_{\max})/2$;

20   Follow step 1 to 12 to obtain $\mathbf{Q}_i^\star$;

21   **if** $\sum_{i=1}^{K} \text{Tr}\{\mathbf{Q}_i^\star\} > P$ **then** set $\lambda_{\min} = \lambda$; **otherwise**, set $\lambda_{\max} = \lambda$

22   **until** $\sum_{i=1}^{K} \text{Tr}\{\mathbf{Q}_i^\star\} = P$ *or* $(\lambda_{\max} - \lambda_{\min})$ *is small enough*;

$$\lambda^\star \sum_{i=1}^{K} \text{Tr}\{\mathbf{Q}_i^\star\} < \lambda^\star \sum_{i=1}^{K} \text{Tr}\{\hat{\mathbf{Q}}_i\}. \tag{7.21}$$

Since Algorithm 7.1 guarantees $\sum_{i=1}^{K} \text{Tr}\{\mathbf{Q}_i^\star\} = P$ for the case $\lambda^\star > 0$, one has

$$P = \sum_{i=1}^{K} \text{Tr}\{\mathbf{Q}_i^\star\} < \sum_{i=1}^{K} \text{Tr}\{\hat{\mathbf{Q}}_i\}, \tag{7.22}$$

which contradicts the condition (i). Thus the proof for this theorem follows by contradiction.

As proved in Theorem 7.1, the optimization (7.8) carried at cell-q can be effectively and optimally solved. For the network WSR maximization problem (7.5), the proposed ILA algorithm requires each cell-$q, q = 1, \dots, Q$ to iteratively update the matrix $\mathbf{A}_q$ and solve its approximated optimization problem (7.8).

Theorem 7.3. *The optimization (7.8) performed at any given BS-q always improves the network WSR $\sum_{q=1}^{Q} \omega_q R_q^{BC}$. Thus, the Gauss-Seidel iterative update is guaranteed to coverage to at least a local maximum.*

Proof. Suppose that $\mathbf{Q}_q = \bar{\mathbf{Q}}_q = \{\bar{\mathbf{Q}}_{qi}\}_{i=1}^{K}, \forall q$ was obtained from the previous iteration, and $\mathbf{Q}_q^{\star} = \{\mathbf{Q}_{qi}^{\star}\}_{i=1}^{K}, \forall q$ is the optimal solution after the current iteration. Note that $f_q(\mathbf{Q}_q, \mathbf{Q}_{-q})$ is a convex function with respect to $\mathbf{Q}_q \in \mathscr{S}_q \triangleq \{\{\mathbf{Q}_{qi}\}|\mathbf{Q}_{qi} \succeq \mathbf{0}, \sum_{i=1}^{K} \mathrm{Tr}\{\mathbf{Q}_{qi}\} \leq P_q\}$ [9, 19]. Thus, the first-order condition of the convex function f_q [1] dictates that

$$f_q(\mathbf{Q}_q^{\star}, \bar{\mathbf{Q}}_{-q}) \geq f_q(\bar{\mathbf{Q}}_q, \bar{\mathbf{Q}}_{-q}) - \sum_{i=1}^{K} \mathrm{Tr}\{\mathbf{A}_q(\mathbf{Q}_{qi}^{\star} - \bar{\mathbf{Q}}_{qi})\} \qquad (7.23)$$

with $\mathbf{A}_q$ being defined in (7.7) at $\bar{\mathbf{Q}}_{qi}$. After one Gauss-Seidel iteration, the network WSR is updated as

$$\sum_{q=1}^{Q} \omega_q R_q^{BC}(\mathbf{Q}_q^{\star}, \bar{\mathbf{Q}}_{-q}) = \omega_q R_q^{BC}(\mathbf{Q}_q^{\star}, \bar{\mathbf{Q}}_{-q}) + f_q(\mathbf{Q}_q^{\star}, \bar{\mathbf{Q}}_{-q})$$

$$\geq \omega_q R_q^{BC}(\mathbf{Q}_q^{\star}, \bar{\mathbf{Q}}_{-q}) + f_q(\bar{\mathbf{Q}}_q, \bar{\mathbf{Q}}_{-q}) - \sum_{i=1}^{K} \mathrm{Tr}\{\mathbf{A}_q(\mathbf{Q}_{qi}^{\star} - \bar{\mathbf{Q}}_{qi})\}$$

$$\geq \omega_q R_q^{BC}(\bar{\mathbf{Q}}_q, \bar{\mathbf{Q}}_{-q}) + f_q(\bar{\mathbf{Q}}_q, \bar{\mathbf{Q}}_{-q}) - \sum_{i=1}^{K} \mathrm{Tr}\{\mathbf{A}_q(\bar{\mathbf{Q}}_{qi} - \bar{\mathbf{Q}}_{qi})\}$$

$$= \sum_{q=1}^{Q} \omega_q R_q^{BC}(\bar{\mathbf{Q}}_q, \bar{\mathbf{Q}}_{-q}),$$

where the first inequality is due to the one in (7.23), and the second inequality is due to $\mathbf{Q}_q^{\star}$ being the optimal solution of problem (7.8). Hence, the network WSR is strictly nondecreasing after an update of the covariance matrices at any given BS. With the Gauss-Seidel (sequential) iterative update, each BS takes turns to refine its precoders and improve the network WSR. Since the network WSR is upper-bounded, the Gauss-Seidel iterative update is guaranteed to converge to at least a local maximum.

The ILA algorithm for the multicell MIMO-BC is summarized in Algorithm 7.2. In Algorithm 7.2, the iterative procedure 2–8 is referred to as an outer-loop iteration and the update at a particular BS using the iterative Algorithm 7.1 in step 6 is referred to as an inner-loop iteration. It is worth mentioning that certain optimization steps in Algorithm 7.2 can be assigned and executed in a distributed manner across the coordinated BSs. We address the distributed implementation of the ILA algorithm in Sect. 7.3.

Algorithm 7.2: ILA Algorithm for the Multicell MIMO-BC with DPC

---

1 Initialize $\{\mathbf{Q}_{q_i}\}_{\forall q, \forall i}$, such that $\sum_{i=1}^{K} \mathrm{Tr}\{\mathbf{Q}_{q_i}\} = P_q$;
2 **repeat**
3      $\tilde{\mathbf{Q}}_{q_i} \leftarrow \mathbf{Q}_{q_i}$;
4   **for** $q = 1, 2, \ldots, Q$ **do**
5   At the BS, update the matrix $\mathbf{A}_q$ as given in (7.7);
6   Update $\mathbf{Q}_{q_i}, i = 1, \ldots, K$ by executing Algorithm 7.1;
7   **end**
8 **until** *convergence*;

7.3 Distributed ILA Algorithm for Multicell MIMO-BC

In order to implement the proposed ILA algorithm distributively, the following assumptions are taken in consideration:

- *Assumption 1*: Each BS, say BS-q, knows the channel matrices $\mathbf{H}_{qr_i}$'s to all the MSs in the network. This assumption allows the BS to control its induced ICI to other cells.
- *Assumption 2*: The coordinated BSs have reliable backhaul links to exchange control information among themselves.
- *Assumption 3*: The channels are in block-fading or vary sufficiently slow such that they can be considered fixed during the optimization process.

It is to be noted that the optimization (7.8) can be performed distributively at the corresponding BS with local information. Thus, it remains to show that the factors $\mathbf{A}_q$'s can also be computed in a distributed manner through a message exchange mechanism among the BSs. It is observed from Eq. (7.7) that in order to compute $\mathbf{A}_q$, BS-q has to possess the channels $\mathbf{H}_{qr_j}$'s to all the MSs in the other cells, as stated in Assumption 1. In addition, BS-q needs to acquire the pricing matrix

$$\mathbf{B}_{r_j} = \omega_r \left[\mathbf{C}_{r_j}^{-1} - \left(\mathbf{C}_{r_j} + \mathbf{H}_{rr_j} \mathbf{Q}_{r_j} \mathbf{H}_{rr_j}^H \right)^{-1} \right] \tag{7.24}$$

from other cells, where $\mathbf{C}_{r_j} = \mathbf{R}_{r_j} + \sum_{k=1}^{j-1} \mathbf{H}_{rr_j} \mathbf{Q}_{r_k} \mathbf{H}_{rr_j}^H$. Thus, it is required that MS-j at cell-r computes its corresponding factor $\mathbf{B}_{r_j}$ using only local measurements. In fact, $\mathbf{C}_{r_j}$ is the total interference plus noise and $\mathbf{C}_{r_j} + \mathbf{H}_{rr_j} \mathbf{Q}_{r_j} \mathbf{H}_{rr_j}^H$ is the total signal and interference plus noise, pertaining to MS-i of cell-q. After computing the pricing matrix $\mathbf{B}_{r_j}$, the MS can feed back this parameter to its connected BS. These factors $\mathbf{B}_{r_j}$'s are then exchanged among the BSs to compute $\mathbf{A}_q$'s.

Remark 4. It is proved in Theorem 7.3 that the optimization carried at a given BS always improves the network WSR, which leads to the convergence of the ILA algorithm with the Gauss-Seidel update. However, before each update, all the MSs are required to compute their pricing matrices $\mathbf{B}_{r_j}$'s and exchange them within the whole network. As a result, the Gauss-Seidel update may demand a lot of computation at the MSs and message exchanges in the network. To reduce the amount of computation and message exchanges, the proposed ILA algorithm can be also implemented by the Jacobi (simultaneous) iterative update. Specifically, after each instance of pricing update and message exchange, all the BSs simultaneously update their covariance matrices. Although the convergence of the ILA algorithm with the Jacobi update is not analytically proved, numerical simulations show a much faster convergence rate by the Jacobi update over the Gauss-Seidel update.

7.4 Simulation Results

This section presents some numerical evaluations on the achievable downlink sum-rate of a multicell system with different levels of coordination and on the convergence behavior of the proposed algorithms. We compare the sum-rate between 3 schemes: (i) the coordination mode with DPC obtained from the ILA algorithm (with equal weights $\omega_1 = \ldots = \omega_Q$), (ii) the competition mode where each BS selfishly maximizes the sum-rate for its connected MSs using DPC, and (iii) the coordination mode with linear precoding obtained from the WMMSE algorithm in [15]. Unless stated otherwise, the ILA scheme is implemented with the Gauss-Seidel update due to its guaranteed convergence. We consider a generic 3-cell system with 3 MSs per cell, as illustrated in Fig. 7.1. The numbers of antennas at each BS and each MS are assumed to be 4 and 2, respectively. The BSs are located at a normalized distance of 2 and the MSs are randomly located on a circle at distance d from its connected BS. The channel coefficients are generated by the path loss model with a path loss exponent of 3. The additive Gaussian noise at each MS is assumed to be white with the covariance matrix $\mathbf{Z}_{q_i} = \sigma^2 \mathbf{I}$ and σ^2 is set at 0.01. The transmit power P_q at each BS is constrained at 1 W, i.e., $P_q/\sigma^2 = 20$ dB, unless stated otherwise.

Figure 7.2 illustrates the total network sum-rate versus the intracell MS-BS distance d obtained from the 3 schemes. As d is varied, 10,000 channel realizations at each value of d are used to obtain the average network sum-rate. Note that the effect of ICI is more apparent with increasing d due to the decreasing gain of intra-cell channels and the increasing gain of inter-cell channels. Thus, the network sum-rate is reduced with increasing d, as observed in the figure for all the schemes. Out of the 3 schemes, scheme (i) always show a superior sum-rate performance, since it takes advantages of both the nonlinear precoding in DPC and the interference coordination. At the low ICI region, i.e, small d, scheme (ii) significantly outperforms scheme (iii) due to the use of DPC over linear precoding. On the other hand, at the high ICI region, by implementing the IC with the proposed algorithm, one can significantly improve the network sum-rate over the competitive design.

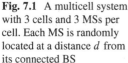

Fig. 7.1 A multicell system
with 3 cells and 3 MSs per
cell. Each MS is randomly
located at a distance d from
its connected BS

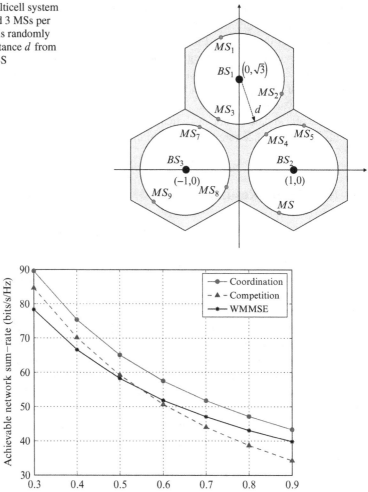

Fig. 7.2 Network sum-rates versus the intra-cell BS-MS distance d

Figure 7.3 illustrates the network sum-rate versus the transmit power to AWGN
ratio P/σ^2 (with same power P at each BS) for $d = 0.7$. It is observed from the
figure that increasing the transmit power at each BS shall increase the network sum-
rate for all 3 schemes. However, at the high P/σ^2 region the sum-rate obtained
from scheme (ii) becomes saturated. This is due to the reason that the competitive
design does not attempt to control ICI, and thus increases the ICI relatively with
the transmit power. In this case, it is more appealing to implement the IC designs in
schemes (i) and (iii). Similar to the observation in Fig. 7.2, the non-linear precoding
design in scheme (i) can extract extra performance from the multicell network over
the linear precoding design in scheme (iii).

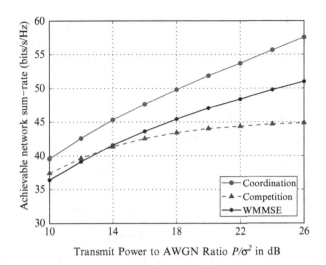

Fig. 7.3 Network sum-rates versus the transmit power to AWGN ratio for $d = 0.7$

It is to be noted that the ILA, WMMSE algorithms, and the competitive design may require many iterations to fully converge. Herein, an iteration is defined as an instance of message exchange among the BSs. As previously mentioned, to implement the interference coordination among the cells, each algorithm (ILA with Gauss-Seidel update and WMMSE) requires the coordinated BSs to exchange signaling messages after each outer-loop iteration. For practical implementation, it may be desirable to set the number of iterations to reduce the amount of message exchanges. Figure 7.4 compares the obtained sum-rates from each algorithm after 10 iterations versus the intra-cell BS-MS distance d. Figure 7.4 shows that the ILA algorithm converges at a faster rate than the WMMSE algorithm in terms of the number of outer-loop iterations. Of the two types of updating in the ILA algorithm, the convergence of the Jacobi update is clearly faster than that of the Gauss-Seidel update.

For a specific channel realization with $d = 0.7$, the convergence behavior of the proposed ILA algorithm is illustrated in Fig. 7.5. After each iteration, the network sum-rates obtained from the algorithms are plotted. As observed from Fig. 7.5, once a BS updates its covariance matrices, the overall network sum-rates are always improved by the ILA algorithm with the Gauss-Seidel update. This behavior agrees with our analysis on the convergence of the algorithms. Interestingly, the ILA algorithm also experiences the monotonic convergence with the Jacobi update. Both types of updating in the ILA algorithm eventually converge to a network sum-rate that is superior than the one obtained by the competitive design.

With the same sample channel realization as in Fig. 7.5, Fig. 7.6 displays the convergence of Algorithm 7.1 in maximizing the BC sum-rate with a penalty term, i.e., problem (7.8). After each update of the dual variable λ, the evolutions of the sum-rate and the transmit power at BS-1 are plotted in the figure. As can be observed

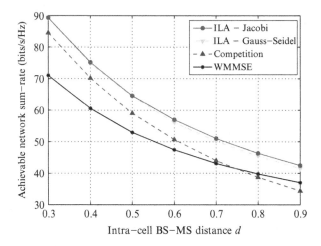

Fig. 7.4 Network sum-rates versus intra-cell BS-MS distance d, obtained from ILA and WMMSE algorithms with 10 outer-loop iterations

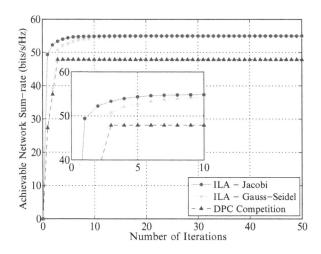

Fig. 7.5 Convergence of the proposed ILA algorithm to maximize the network sum-rate with interference coordination

from the figure, the algorithm converges very fast in a few iterations. Due to the penalty term, BS-1 needs to *balance* its achievable sum-rate with the ICI induced to cell-2 and -3. Thus, its sum-rate is undoubtedly reduced, compared to the one obtained in a conventional BC without the penalty term. Nonetheless, under the IC mode, each BS adopts a less selfish strategy to improve the overall network sum-rate, as shown in Fig. 7.5.

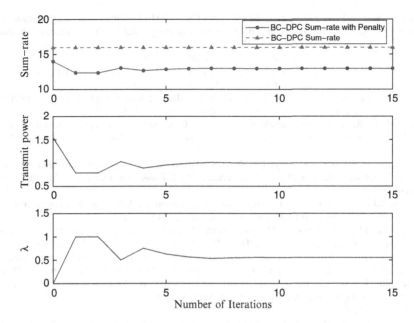

Fig. 7.6 The convergence of Algorithm 7.1 to solve Problem (7.8)

7.5 Concluding Remarks

This chapter examined the problem of network WSR maximization in the multi-cell MIMO-BC with DPC. Under the coordination mode, the network sum-rate maximization problem was shown to be nonconvex. This work considered a low-complexity solution approaches, namely ILA, to search for a locally optimal solution. In the first approach, successive convex approximation technique was utilized to transform the problem into multiple per-cell problems, which are then optimized distributively at each BS. In particular, each BS attempted to maximize the BC sum-rate to its connected BS with a penalty term on its induced ICI to other cells. A distributed and fast converging algorithm was then proposed to efficiently find a locally optimal solution to the network WSR maximization problem. As the proposed algorithms allow the multicell to take advantage of both DPC and coordinating the ICI, they show a significant improvement in the network sum-rate compared to competitive design and the linear precoding.

References

1. Boyd, S., Vandenberghe, L.: Convex Optimization. Cambridge University Press, United Kingdom (2004)
2. Caire, G., Shamai, S.: On the achievable throughput of a multiantenna Gaussian broadcast channel. IEEE Trans. Inform. Theory **49**(7), 1691–1706 (2003)

3. Christensen, S.S., Argawal, R., E. de Carvalho, Cioffi, J.M.: Weighted sum-rate maximization using weighted MMSE for MIMO-BC beamforming design. IEEE Trans. Wireless Commun. **7**(12), 4792—4799 (2008)
4. Costa, M.: Writing on dirty paper. IEEE Trans. Inform. Theory **29**(3), 439–441 (1983)
5. Dahrouj, H., Yu, W.: Coordinated beamforming for the multicell multi-antenna wireless system. IEEE Trans. Wireless Commun. **9**(5), 1748–1759 (2010)
6. Huh, H., Papadopoulos, H.C., Caire, G.: Multiuser MISO transmitter optimization for intercell interference mitigation. IEEE Trans. Signal Process. **58**(8), 4272–4285 (2010)
7. Jindal, N., Rhee, W., Vishwanath, S., Jafar, S.A., Goldsmith, A.: Sum power iterative water-filling for multi-antenna Gaussian broadcast channels. IEEE Trans. Inform. Theory **51**(4), 1570–1580 (2005)
8. Kaviani, S., Simeone, O., Krzymień, W.A., S. Shamai (Shitz): Linear precoding and equalization for network MIMO with partial cooperation. IEEE Trans. Veh. Technol. **61**(5), 2083–2096 (2012)
9. Kim, S.J., Giannakis, G.B.: Optimal resource allocation for MIMO ad hoc cognitive radio networks. IEEE Trans. Inform. Theory **57**(5), 3117–3131 (2011)
10. Nguyen, D.H.N., Le-Ngoc, T.: Multiuser downlink beamforming in multicell wireless systems: A game theoretical approach. IEEE Trans. Signal Process. **59**(7), 3326–3338 (2011)
11. Nguyen, D.H.N., Le-Ngoc, T.: Sum-rate maximization in the multicell MIMO multiple-access channel with interference coordination. IEEE Trans. Wireless Commun. **13**(1), 36–48 (2014)
12. Schmidt, D., Shi, C., Berry, R.A., Honig, M.L., Honig, W.: Minimum mean square error interference alignment. In: Assilomar Conf. Signals, Syst. and Comp., pp. 1106–1110. Pacific Grove, CA (2009)
13. Scutari, G., Palomar, D.P., Barbarossa, S.: Competitive design of multiuser MIMO system based on game theory: a unified view. IEEE J. Select. Areas in Commun. **26**(9), 1089–1102 (2008)
14. Shi, C., Berry, R.A., Honig, M.L.: Monotonic convergence of distributed interference pricing in wireless networks. In: Proc. IEEE Int. Symp. Inform. Theory, pp. 1619–1623. Seoul, Republic of Korea (2009)
15. Shi, Q., Razaviyayn, M., Luo, Z.Q., He, C.: An iteratively weighted MMSE approach to distributed sum-utility maximization for MIMO interfering broadcast channel. IEEE Trans. Signal Process. **59**(9), 4331–4340 (2011)
16. Venturino, L., Prasad, N., Wang, X.: Coordinated linear beamforming in downlink multi-cell wireless networks. IEEE Trans. Wireless Commun. **9**(4), 1451—1461 (2010)
17. Vishwanath, S., Jindal, N., Goldsmith, A.: Duality, achievable rates and sum-rate capacity of Gaussian MIMO broadcast channels. IEEE Trans. Inform. Theory **49**(10), 2658–2668 (2003)
18. Viswanath, P., Tse, D.: Sum capacity of the vector Gaussian broadcast channel and uplink-downlink duality. IEEE Trans. Inform. Theory **49**(8), 1912–1921 (2003)
19. Ye, S., Blum, R.S.: Optimized signaling for MIMO interference systems with feedback. IEEE Trans. Signal Process. **51**(11), 2839–2848 (2003)
20. Yu, W., Cioffi, J.: Sum capacity of Gaussian vector broadcast channels. IEEE Trans. Inform. Theory **50**(9), 1875–1892 (2004)
21. Yu, W., Lan, T.: Transmitter optimization for the multi-antenna downlink with per-antenna power constraints. IEEE Trans. Signal Process. **55**(6), 2646–2660 (2007)
22. Zhang, L., Zhang, R., Liang, Y.C., Xin, Y., Poor, H.V.: On Gaussian MIMO BC-MAC duality with multiple transmit covariance constraints. IEEE Trans. Inform. Theory **58**(4), 2064–2078 (2012)